FIREFIGHTER

Haynes®

(All roles and skills)

COVER IMAGE: Firefighters are provided with uniforms and equipment which affords them high levels of personal protection. *(Ian Moores)*

First published in October 2019

A catalogue record for this book is available from the British Library.

ISBN 978 1 78521 205 5

Library of Congress control no. 2018953077

Published by Haynes Publishing,
Sparkford, Yeovil, Somerset BA22 7JJ, UK.
Tel: 01963 440635
Int. tel: +44 1963 440635
Website: www.haynes.com

Haynes North America Inc.,
859 Lawrence Drive, Newbury Park,
California 91320, USA.

Printed in Malaysia.

FIREFIGHTER

(All roles and skills)

Owners' Workshop Manual

An insight into the training, equipment, roles and working lives of firefighters

Philip Martin & Duncan J. White

Contents

6 Introduction

8 The firefighter story

The outbreak of the Second World War 11
The post-war fire service 12

14 The modern firefighter – what makes a good firefighter?

Joining the fire and rescue service 16
Duty systems 17
Roles 18
Fire control staff 21
Defence Fire Risk Management Organisation 21

24 Delivering a professional fire and rescue service

Integrated risk management plan 26
Training 26
Theoretical knowledge 29
Becoming a competent firefighter 36
Fire and rescue service activities 39
Response 50

56 The fire station

Internal facilities in the fire station 58
External facilities in the fire station 61
Combined emergency service stations 62
Community use of fire stations 62

64 Equipment

Personal protective equipment 66
Breathing apparatus entry control procedures 71
Thermal imaging cameras 73
Firefighting equipment 73
Water supplies 78
Firefighting hose 79
Very high-pressure cutting and extinguishing systems 84
Firefighting foam 84
Portable fire extinguishers 88
Positive pressure ventilation 90
Rescue and extrication equipment 91
Stabilisation 95
Glass management 96
Airbag cover 98
Small gear 98
Emergency medical equipment 98

LEFT Firefighters battle a large fire involving over 150 caravans at a caravanning sales site. The loss from this fire was estimated at over £1 million. *(Paul Chamberlain)*

OPPOSITE TOP Firefighters use compressed air foam jets as they work to save the thatch roof of a public house in Somerset, which was originally built in the 15th Century. *(Paul Chamberlain)*

100 Fire appliances

Major rescue pump 103
Light rescue pumps 104
Rapid intervention vehicles 105
Off-road fire capability 106
High reach extendable turret 106
Airport rescue and firefighting appliances 107
Aerial ladder platform 108
Specialist appliances 110
Urban search and rescue 115
High-volume pumping 116
Mass decontamination 117
External support vehicles 118

120 The anatomy of an emergency call

Calling 999 122
Fire and rescue service control room 122
Arrival on scene 123
Fire control rooms 125
Fire and Rescue Service National Control Centre 125

126 Emergency incidents

Fires 128
Rural firefighting 136
Technical rescue 138
Other specialist operations 142

144 Major incident case studies

Buncefield Fuel Depot explosion and fire 146
Boscastle flood 148
7/7 London terrorist attacks 150
M5 motorway collision and fire 150
Grenfell Tower fire 151

152 Fire investigation

156 Firefighter welfare – post-trauma prevention and support

160 Future firefighting

Predictive analytics 162
Electric-powered fire appliances 162
Nanotechnology 162
Drones 162
Radio frequency identification tracking 163
Powered exoskeletons 163

164 Appendices

Support services 164
Model recruitment guidance 166
Glossary of fire and rescue service acronyms 168

170 Index

Introduction

Firefighters share a special place in the hearts of the public. Many a young child will have firefighter at the top of their list of what they would like to do when they grow up. Even as adults, we are somehow drawn to watch as a fire engine passes us in the street or on the road.

It seems that part of the appeal of firefighters is their desire to help anyone in need without question. The image of the firefighter is often that of them selflessly entering a burning building without a second thought for their own safety. Over history, whether through art or news coverage, the firefighter is often depicted as the character carrying their helpless victim to safety.

The modern firefighter has a very wide and diverse role, requiring them to learn a wide range of skills, to know how to operate an array of often complex tools and machinery and to understand the science and mechanics behind fires and other incidents.

To many of us, I am sure that the emergency role of the firefighter is what we consider as their main activity, and while this role remains incredibly important, it is just one aspect of a much wider role in what remains a respected profession.

To understand the wider role there is a need to look beyond the very visible response role of the firefighter and explore the variety of activities performed by the men and women of the modern fire and rescue service. In order to understand the expectations of society on the modern firefighter it is also important to understand some of the history behind the very origins of firefighting through time to the modern role.

The next step is to follow the journey taken by a newly appointed firefighter to explore the content and challenges of basic training, the range of skills and knowledge that needs to be gained before they can attend an emergency, and the breadth of the fire and rescue service role.

Understanding the science behind a fire is critical for effective firefighting. Looking at the fundamental requirements for a fire to burn provides a background to understanding how fire is then extinguished.

Beyond the initial training and formation of skilled teams is the need to provide a 24-hour emergency service. This is achieved through a range of duty systems operated to meet the risks and demands. The structure of the fire and rescue service is built on a range of roles managing and commanding operations, and a career path exists to support this. Beyond the operational firefighter role, it is important to understand the critical part that the fire control room and its teams perform in ensuring the right resources are quickly sent to and supported at an incident.

With progress in technology and the recognition of the benefit and importance of safety education for society, firefighters have been recognised for their success in not only responding to but also in preventing emergencies and protecting individuals and communities when an incident does occur.

Recognising that the fire and rescue service is just one of the emergency services and that working together brings better results, each of the UK emergency services share a common operating protocol that ensures a common and effective response to incidents where more than one emergency service is required.

As a team, firefighters operate out of fire stations across the country. To many, the inside of a modern fire station remains a mystery.

The design, accommodation and facilities provided are all key to ensuring firefighters can maintain their skills and equipment, with the ability to respond with the right equipment at a moment's notice.

Fire engines, or fire appliances as they are known in the service, come in many shapes and sizes with a wide capability created by their design and the equipment carried. The capability of the fire and rescue service is enhanced significantly by the range of equipment that is available.

A response to any incident has to come from an initial call. The journey from call to attendance and the interactions and technology that make it happen so effectively and efficiently demands well-trained staff with the right skills and knowledge.

The range of incidents the fire and rescue service attends covers the whole of the built environment and the transport systems used to move people and goods. Each can present its own unique challenges. Using incidents of note, it is possible to examine the response to a number of high-profile but diverse incidents that have occurred in recent history.

Understanding what went wrong is key to learning and getting things right in the future is at the heart of the role of fire investigation. In addition, the investigation of a fire scene is key in supporting the role of the police where a crime is suspected and of Her Majesty's Coroner when a life is lost.

With technology key to finding and implementing solutions to current and future issues, this book will conclude by considering some of the emerging opportunities for the future of fire prevention and firefighting presented by innovation and technology.

ABOVE Firefighters tackling a fire in the roof of a domestic house. *(Cornwall FRS)*

BELOW Fires in commercial premises often require the use of specialist appliances. *(Cornwall FRS, Bude)*

Kirkwall
Stornoway
The Minch
Loch Shin
North Uist
South Uist
Skye
Inverness
Elgin
Loch Ness
Aberdeen
ATLANTIC
OCEAN
Loch Rannach
Mull
Oban
SCOTLAND
Loch Tay
Forfar
Arbroath
Dundee
Earn
Jura
Lochgilphead
Stirling
Loch Lamond
Islay
Kirkcaldy
Greenock
Edinburgh
Haddington
Glasgow
Motherwell
Dalkeith
Gigha
Irvine
Kilmarnock
Berwick-upon-Tweed
NORTH
SEA
Ayr
Arran
North Channel
Coleraine
Londonderry
Hawick
Nith
Strabane
Ballymena
Dumfries
Larne
Omagh
Esk
NORTHERN IRELAND
Tynemouth
South Shields
Enniskillen
Bangor
Carlisle
Newcastle
Sunderland
Belfast
Workington
Durham
Whitehaven
Armagh
Wear
Hartlepool
Tees
Stockton-on-Tees
Newry
Middlesborough
Darlington
Isle of Man
Dundalk
Douglas
Ure
Barrow-in-Furness
Scarborough
Morecambe
Lancester
Harrogate
IRISH SEA
York
Blackpool
Blackburn
Bradford
Leeds
Halifax
Preston
DUBLIN
Kingston upon Hull
Bolton
Anglesey
Manchester
Liverpool
Wigan
Scunthorpe
Barnsley
IRELAND
Birkenhead
Warrington
Doncaster
Great Grimsby
Colwyn
Bay
Stockport
Rotterham
Trent
Caernarfon
Chester
Macclesfield
Sheffield
Chesterfield
Crewe
Wrexham
Stoke-
on-Trent
Lincoln
Mansfield
Notthingham
Derby
Boston
The
Wash
Shrewsbury
Stafford
Cardigan
Grantham
Waterford
Telford
Burton-
upon-Trent
Loughborough
Bay
Wolverhampton
Walsall
Leicester
Aberystwyth
King's Lynn
St Georges Channel
Birmingham
Coventry
Norwich
Llandrindod
Wells
Kidderminster
Rugby
Peterborough
Great Yarmout
WALES
Worchester
Warwick
Lowestoft
Kettering
Hereford
ENGLAND
Wye
Northampton
Bury St Edmunds
Haverfordwest
Carmarthen
Cambridge
Ipswich
Cheltenham
Milton Keynes
Swansea
Gloucester
Aylesbury
Colchester
Oxford
Luton
Cardiff
Newport
Swindon
St Albans
Chelmsford
Bristol Channel
CELTIC SEA
Bristol
LONDON
Thames
Weston-Super-Mare
Southend-on-Sea
Bath
Barnstaple
Trowbridge
Newbury
Reading
Kingston
Gillingham
Margate
Basingstoke
Ramsgate
Exe
Guildford
Maidstone
Canterbury
Yeovil
Salisbury
Crawley
Royal
Tunbridge Weels
Dover
Exeter
Southampton
Dorchester
Portsmouth
Folkestone
Strait of Dover
Chichester
Worthing
Weymouth
Bournemouth
Hastings
Calais
Torquay
Newport
Brighton
Eastbourne
Isle of Wight
Isles of Scilly
Penzance
Plymouth
Falmouth
ENGLISH CHANNEL
FRANCE

RESCUE

Chapter Two

The modern firefighter – what makes a good firefighter?

Firefighters are drawn from all parts of our community. The emergency response aspect of the role requires firefighters to be physically fit, be able to solve problems and work with initiative.

OPPOSITE The ability to work together as part of a team is the cornerstone to a successful career in the fire and rescue service. *(Shutterstock)*

RIGHT Fire safety education and prevention activities form a major part of a firefighters role. *(Cornwall FRS)*

Good communication skills are essential for effective teamwork and when dealing with members of the public in potentially confusing and distressing environments. Firefighters are required to bring calm in the midst of chaos and be caring when there is confusion.

Behaving with and treating others with dignity and respect, regardless of background or circumstances, is essential for firefighters. In order to be effective in their role, firefighters have to earn and hold the respect of the people they meet in all areas of their work.

Firefighters spend a large part of their time working with people to help them live safer lives. These people range from the very young in schools, through working with businesses to ensure they operate places that are safe to work, live and visit, and working with older people who need to be safe in their homes.

BELOW, BOTTOM AND BOTTOM RIGHT Modern fire stations vary in size, dependent on the communities and risks they cover. *(Cornwall FRS)*

Joining the Fire and Rescue Service

The role of a firefighter is a popular career choice. Many fire and rescue services report that when vacancies are advertised they are oversubscribed, giving services the opportunity to select only the very best.

If you are considering becoming a firefighter, there are different opportunities depending on the particular role you are seeking. Opportunities exist for full-time and on-call operational firefighters and for the role of firefighter (control). The latter involves answering emergency calls and mobilising resources to incidents.

ABOVE **Multiple extension ladders pitched at the scene of a fire in a row of terrace properties, allowing firefighters greater access to the roof space.**
(Steve Greenaway)

Duty systems

The duty system at any particular fire station will be dependent on the nature of the risks and the frequency of incidents in a particular area. Areas of highest risk with large numbers of calls are likely to be staffed 24 hours a day, whereas quieter areas with lower risk will be covered by a system where firefighters respond to the station only when there is an incident. The types of duty system most commonly used are explained below.

Wholetime

A wholetime fire station is one where at least one fire appliance crew is on duty 24/7. Firefighters work in shifts and spend that shift with their fire appliance and crew members to ensure the quickest response is achieved. Firefighters are not confined to the fire station and will spend large parts of their respective shifts training or carrying out work to reduce risks in the communities they serve.

Day-crewed

A day-crewed fire station will have a mix of firefighters on different shift systems. During the busiest period there will be a crew staffing at least one fire engine similar to that of a wholetime station. In quieter periods, the duty system at the fire station will change, with on-call firefighters providing cover.

On-call

An on-call (retained) duty system involves firefighters attending the fire station in the event of a fire or for training or other activities. On-call firefighters carry a pager, which is operated by the fire and rescue service control centre when there is an incident in their area. The firefighters then respond to the fire station and on to the incident in their fire appliance.

Mixed duty systems

Many fire stations operate two or more different duty systems. This provides flexibility, enabling the appropriate numbers of firefighters at any given time. For example, a wholetime fire station may have three or four fire appliances with a permanent crew of just five firefighters. On-call firefighters are also available when required. This enables the wholetime crew to staff different fire appliances as required and for additional firefighters to supplement in the event of a larger incident or when the wholetime crew is already committed to another incident.

Flexible duty system for supervisory managers

The majority of incidents attended by fire crews can be dealt with by one or two fire engines. At larger incidents there may be a need to have a number of additional managers to deal with larger numbers of firefighters or a manager experienced in specialist roles such

ABOVE Fire Officers respond to incidents in specially designed vehicles. *(Cornwall FRS/Michael Soanes)*

BELOW Volunteer crew at Leadhills Community Fire Station, Scotland. *(Scottish FRS)*

BOTTOM Apprentice firefighters undertaking basic training, under the watchful eye of their instructors. *(Cornwall FRS)*

as hazardous materials, fire investigation or technical rescue.

These specialist officers will generally have a managerial responsibility that forms their general role. They may be managing a number of fire stations or managing other areas of the service, such as training, fire protection or community safety, and will primarily work from an office in the normal working day, being available to provide cover from their office, home or another base.

Officers working the flexible duty system provide standby cover outside of their normal working hours. In the event of them being needed at an incident they will be called by fire control via a mobile phone or a paging device. They will then respond to the incident in a car or smaller vehicle. When such an officer is required, the fire control room will contact them to mobilise them.

Volunteers

Some fire and rescue services use volunteer firefighters. This will usually be where the demand for an emergency response is limited but the local community support a system of local cover.

In most circumstances the volunteers will be overseen and mobilised by the local fire and rescue service. Volunteer firefighters will often complete their training and preparation activities in their own time, receiving payment only when mobilised to an incident by their overseeing service.

Volunteer firefighters are required to complete training to the same standard as their paid colleagues. They will have training that matches the equipment that they have and the risk that they could be expected to confront.

Apprentices

Some fire and rescue services operate an apprentice scheme for young persons who wish to explore a career in the fire service. There are varying arrangements for volunteer firefighters with training based on that of professional firefighters.

Roles

Within the fire service there are different roles, depending on managerial responsibility and the expectations of the individual. These roles are explored as follows.

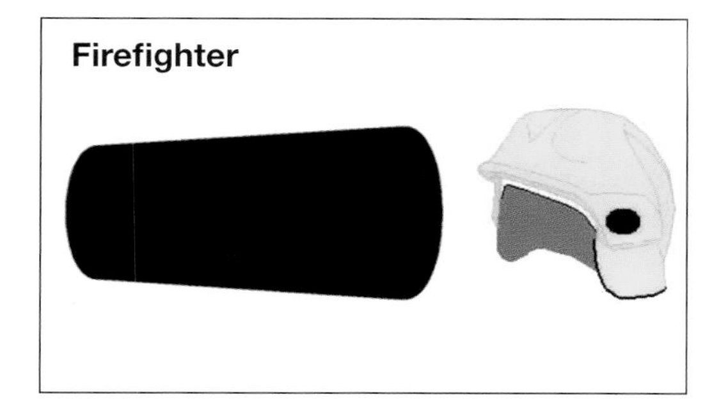

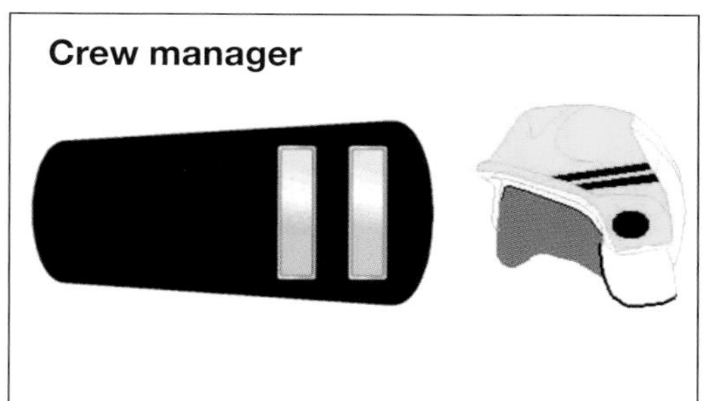

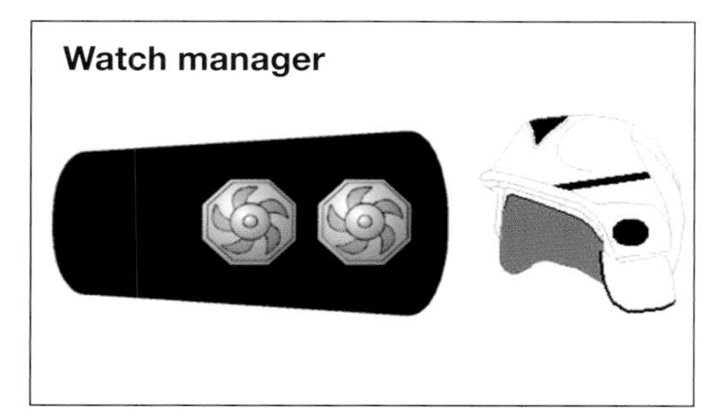

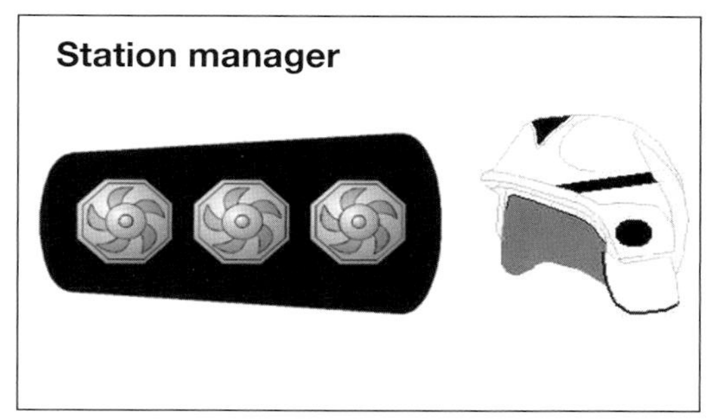

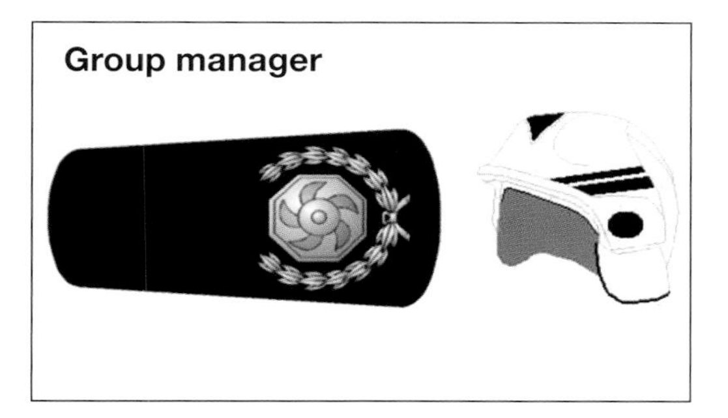

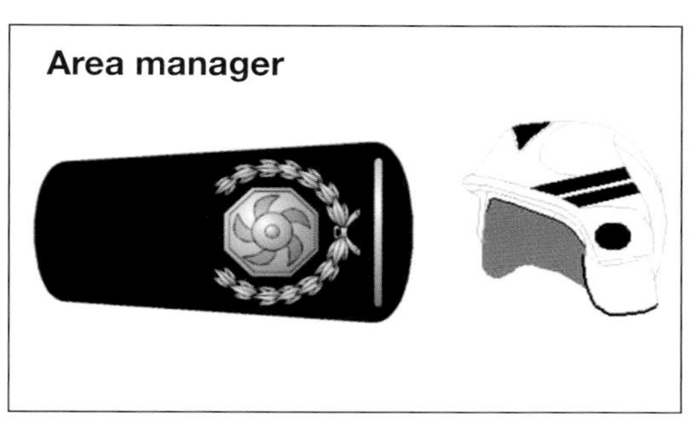

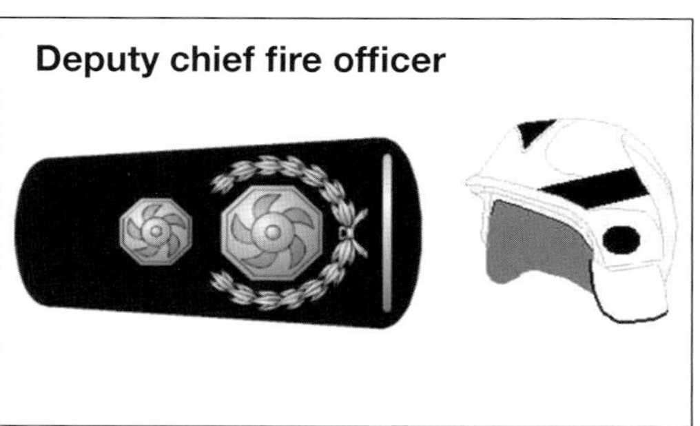

THIS PAGE
Roles within the fire and rescue service are identified by the use of badges and helmet markings.
(Cumbria FRS)

Firefighter

The role of firefighter is the entry level for an operational career. Many firefighters choose to stay in the role for their entire career. Firefighters have opportunities to develop within their role and will gain specialist skills such as the wearing of breathing apparatus, the use of road traffic rescue equipment, first aid, driving and other skills commensurate with their role.

Firefighters primarily act as an operational crew member. Where required, firefighters may perform specialist roles where they have specific skills in those areas.

At an incident, firefighters will be the key to the physical role of firefighting and rescue. The team of firefighters at an incident operate with a supervisor who will be responsible for overseeing the tasks they are completing.

Progression between roles requires individuals to complete an assessment centre and interview, which will include a test of their technical knowledge.

Crew Manager

The first level of uniformed manager in the service is that of crew manager. A crew manager who responds to incidents forms part of an operational crew. The crew manager can command smaller incidents or supervise a team or task at a larger incident.

Crew managers may also carry out specialist roles where the type of work they are required to complete demands additional levels of responsibility, for example community safety work, training, research and other support roles.

BELOW A firefighter (Control) is responsible for call handling and dispatching fire appliances. *(Cornwall FRS/Caterina Lombardi)*

Watch Manager

As the second level of uniformed manager, a watch manager holds responsibility for the management of the team that forms a watch (the term commonly used for a group of firefighters who regularly work together on a shift).

At an incident, a watch manager may take command and be responsible for a team of firefighters, or a number of teams who are under the command of a crew manager.

Watch managers may be appointed into specialist and support roles, for example as instructors, fire protection enforcement officers, training managers and various other specialist roles.

Station Manager

A station manager may have responsibility for the management of one or a number of fire stations. In this capacity the watch or crew managers will report to him. At an incident, the station manager can be responsible for larger or specialist incidents and may have a specialist advisory role.

A station manager may hold responsibility for a team or teams within a specialist role such as fire safety, training, operational planning or a specific project.

Group Manager

A group manager will have responsibility for a group of fire stations (often within a geographical area) or a department of their service such as training, community safety or fire protection. They may also lead a particular project within the service.

At an incident, a group manager will be responsible for larger or specialist incidents and may have a specialist advisory role.

Area Manager

An area manager may have responsibility for a number of groups in an operational geographic area or a number of specialist departments.

An area manager performs a strategic management role and may attend a large incident to take command, or perform an incident management role remote from the incident where a higher command structure has been established.

Brigade Manager

The brigade manager role will hold responsibility for the day-to-day strategic management of the service and will include brigade managers who are assistant chief fire officers, assistant fire commissioners, deputy chief fire officers, deputy fire commissioners, chief fire officers and fire commissioners. Brigade Managers are ultimately responsible to the public, the fire authority and the government for ensuring that their service is managed and delivered efficiently and effectively.

Fire control staff

Fire control rooms will often operate similar shift patterns to fire stations. Fire control rooms are staffed for 24 hours a day. The number of staff on a fire control watch will be heavily dependent on the demands made to that particular control room. Each fire control watch will have a supervisor or supervisors commensurate with a number of people in the team. Fire control teams have a similar role to operational responding crews, as explained below.

Firefighter (Control)

Within the fire control room a firefighter (control) performs the role of call handler and dispatcher. This involves receiving emergency calls and mobilising fire appliances and crews, communicating with crews at an incident and responding to requests for assistance. Responsibilities also include liaising with other emergency services and specialist resources and ensuring the maintenance of cover across areas where resources may be reduced due to ongoing incidents.

Crew Manager (Control)

Crew managers within a fire control are responsible for the supervision of a number of operators. This requires supervision of mobilising and communications with fire appliances and crews and ensuring the maintenance of cover.

Watch Manager (Control)

The watch managers within a fire control may be responsible for a watch or number of watches. A watch manager will supervise operators and crew managers. A crew or watch manager in charge of fire control will be responsible for the correct response and communication with incident commanders and the management of the fire control team on duty.

Station Manager (Control)

The station manager (control) takes responsibility for the control room and the supervisory managers. They may be required to work as a supervisor in the control room where operational activity demands.

Group Manager (Control)

The group manager (control) is likely to hold overall responsibility for the control room function, equipment within and relationships with technical support teams and operational managers. They often hold responsibility for the control room policies and procedure, as well as infrastructure (computers, mobilising systems, radio and telephone communications) and the system's management and lifecycle.

Defence Fire Risk Management Organisation

There are specific fire and rescue arrangements for many UK-based sites operated by the Ministry of Defence. These arrangements fall under the authority of the Defence Fire Risk Management Organisation (DFRMO).

BELOW DFRMO firefighters protect the assets of the Ministry of Defence. *(OGL)*

DFRMO provides fire and rescue cover and response for the protection across buildings and strategic sites and equipment within over 4,600 sites, including 45,000 buildings and 300,000 Ministry of Defence employees.

Civilian firefighters who form the Defence Fire and Rescue Service provide cover for the Ministry of Defence estate and assets.

Royal Air Force firefighters

Royal Air Force (RAF) firefighters are trained to meet defence requirements, therefore also meeting the National Occupational Standards for Local Authority Fire Services. This provides RAF firefighters with the capability to provide a competent self-protecting fire and rescue response to a range of incidents.

BELOW A Royal Air force firefighter undertaking training on an aircraft fire training simulator. *(OGL)*

RAF firefighters provide 24/7 fire and crash rescue protection, with crews organised and trained to respond to any emergency with highly specialised equipment.

Apart from aircraft rescue and firefighting (ARFF), RAF firefighters are also trained in conventional firefighting, enabling them to respond to and deal with fires in technical areas and living quarters. RAF firefighters assist their local authority fire and rescue colleagues in local off-base incidents.

The RAF fire and rescue service has two specialised units, providing fire cover to special safety team convoys on the UK roads and to The Queen's Flight.

In April 2005, following a project known as Fire Study 2005, it was formally agreed that the new model for Fire Services in Defence would be created.

In 2006 the changes took effect and the DFRMO was created at Abingdon to manage all MOD fire services worldwide, including RAF firefighters and Royal Navy aircraft handlers.

RAF firefighters have served all over the world from Africa to the Middle East, India, Far East, Central America, Falklands, Canada and Europe, and have been required to operate in hostile environments and conflict zones, which always bring new challenges.

RAF firefighters are classified as dismounted close combat troops and form an integral element of any RAF force protection capability. As military personnel, RAF firefighters provide their own force protection and are trained to fight fire while under fire. Therefore, RAF firefighters can operate semi-autonomously without the need for extensive external force protection or fire and rescue support and as part of a coalition force, sharing vehicles and equipment with partner nations.

In Iraq and Afghanistan, RAF fire service personnel responded to incidents in hostile situations in support of medical emergency response teams, contributing to saving lives in the most difficult of circumstances.

RAF firefighters are also employed within fire safety roles throughout the world. UK Defence is currently operating in around 40 countries. Many operations are small deployments of personnel employed in Defence Engagement; working with organisations such as NATO and the UN to develop militaries in third world nations.

LEFT **Royal Navy firefighters face unique challenges when firefighting inside the hull of a ship.** *(OGL)*

Royal Navy firefighters

Every Royal Navy sailor is trained to fight fires to prepare them for service onboard the Royal Navy's ships and submarines. Royal Navy fire training specialises in ship compartment firefighting, including areas such as machinery control rooms, engine rooms, galleys, ammunition stores, passageways and accommodation areas.

Aircraft handlers are a branch of the Royal Navy's Fleet Air Arm who specialise in aircraft fire and rescue as well as aircraft hangar firefighting while on board ships that carry aircraft.

Members of the Naval Air Command Fire & Rescue Service also provide the aircraft crash rescue firefighting duties and fire safety at Royal Naval Air Stations.

Works fire services

Some large commercial and industrial complexes employ their own fire teams in both full-time and part-time capacities. Equipment, standards, training and structures are generally based on those of the wider local authority service, although staff will have enhanced knowledge and skills, based on the type of risk they are covering.

STRUCTURE OF THE FIRE AND RESCUE SERVICE

The governance structures for local authority fire and rescue services in the United Kingdom are varied. They are established and operate as follows:

- **England and Wales** – Fire and Rescue Services Act of 2004
- **Scotland** – Fire Scotland Act 2005 (as amended in the Police and Fire Reform (Scotland) Act 2012)
- **Northern Ireland** – The Fire and Rescue Services (Northern Ireland) Order 2006

BELOW **A fire appliance used by an Industrial FRS – it is often just the crest on the side of the appliance which distinguishes it from a regular Fire and Rescue Service appliance.** *(Emergency One)*

Chapter Three

Delivering a professional fire and rescue service

The fire and rescue service undertakes and delivers a range of activities. In this chapter we will explore how the service identifies and responds to risks in the community, trains their staff, educates the public, ensures buildings are safe, responds to a wide range of emergencies and ensures their staff are readily prepared for all eventualities.

OPPOSITE Realistic training is vital to prepare firefighters for their role. *(Cornwall FRS)*

ABOVE **The IRMP of Merseyside Fire and Rescue Authority.** *(Merseyside FRS)*

Integrated Risk Management Plan

In order to be prepared, each fire and rescue authority is required to complete an integrated risk management plan (IRMP) for their area. The IRMP considers the types of risk in the community and informs the fire and rescue authority to ensure the fire and rescue service (FRS) has the right resources in the right place at the right time to deal with the risks.

Fire and rescue services established their first IRMP following national guidance in 2003. This approach represented a significant shift from the previous system of fire cover standards that had been nationally prescribed. This change coincided with a change in the emphasis of governance with the UK fire and rescue services, the principle contending that each fire and rescue service is different and should therefore be able to direct its resources as appropriate for the risk that exists.

In creating and maintaining an up-to-date IRMP, fire and rescue services will gather information to create a risk profile for the service area. This includes information on transport networks, commercial premises, industrial risk, planned changes to infrastructure, new buildings, home safety risks, living accommodation risks and demographic data, which details human factors such as age profiles, social influences, types of accommodation and health data.

Historical data from within both the service and through other organisations such as the health service, councils and other emergency services is also used to help understand and predict what influences and constitutes local risk.

Once the risk profile has been established, the IRMP process helps to determine and plan for the most appropriate means of reducing and responding to risk. From this, the fire and rescue service will target resources through activities such as home safety visits, community education programmes and fire protection activities in workplaces and other commercial premises. Response resources are also evaluated to ensure they remain the most appropriate and in the best location.

The IRMP are required to cover a minimum of a three-year period. It prompts a review of resources and may lead to moving fire appliances and fire engines where risk has changed, increased or reduced.

Training

Recognising that the role of the FRS is wide and varied, firefighters have to develop the skills necessary for carrying out work to prevent incidents, ensure the safety of people in buildings and respond to a wide range of emergencies. There are specialist areas of skill and responsibility, but all operational firefighters are required to attain, develop and uphold core skills.

Initial skills training

Basic firefighting techniques

Initial training for firefighters on appointment is delivered through specialist trainers as part of a basic training programme. The timeframe and approach for delivery of this training will vary, dependent on the duty system, but the competences required will be similar for all and these are based on agreed national occupational standards.

Following recruitment, a newly appointed firefighter will attend an initial training course. This will equip the firefighter with basic fire and rescue knowledge, skills and competencies through a programme that has a mix of practical and theoretical learning.

LEFT Firefighters preparing to undertake a standard drill. *(Cornwall FRS)*

A successful element of initial firefighting training involves carrying out standard drills. With their basis in military training, standard drills require participants to follow clear instructions to achieve an end objective. Through practice and understanding, each member of a crew carrying out a drill will systematically follow instructions so that the individual contribution leads to the whole team meeting the end objective.

At the start of a standard drill, each crew member is allocated a number relating to their position and role in the drill. The instructor will direct each member of the team to carry out certain tasks. This is an autocratic approach, where deviation from the instruction would affect the overall team performance. The result is an orchestrated team effort that can be replicated to consistently achieve the same outcome.

This approach has been a proven method of instilling confidence, teamwork and practical skills for many years. After theoretical input and slow-time practices, trainee firefighters are able to consistently practise these skills, moving to a different position in order to gain a rounded experience of all of the individual task elements.

Standard drills do have their limitations and are designed to develop the basic skills for a range of tasks. As firefighters develop, the experience gained in drills underpins their ability to broaden their capability and to operate more intuitively beyond rigid boundaries.

Water supplies

As part of their basic skills training, firefighters must understand the tools and equipment that will provide them with access to water supplies beyond those carried in their fire appliance. This will include using open water sources, such as rivers, streams and ponds, as well as purpose-built open water supplies and fire hydrants.

When pumping from open water, firefighters will learn how a power-driven pump creates an initial vacuum to enable a pump to provide

LEFT Firefighters preparing to pump from an open water supply using a light portable pump. *(Cornwall FRS)*

RIGHT A fire hydrant in use at an incident to provide a water supply to the fire appliance. *(Cornwall FRS)*

FIRE FACT

Prior to the amalgamation of fire services into the National Fire Service in 1947, the connections on hose lines varied across different and sometimes neighbouring services. The hose connections used by fire and rescue services in the UK today remain of the design adopted in 1947.

energy to a water supply before sending it through fixed pipework and into firefighting hoses. They must understand the impact of atmospheric pressure, the hydraulic impact of a rise in height that will either increase the energy required to create a vacuum or reduce the pressure as water is pumped higher.

In urban and semi-urban areas, the most likely water supply will be from fire hydrants, which draw water from water mains through hydrants set beneath pavements and roads. Firefighters will learn how water mains are configured and how different types of fire hydrant operate. They will learn how to fix standpipes that take a water supply from the hydrant outlet into fire hoses and how to operate the valve safely. Improper use of a fire hydrant can cause issues with both the water mains and the quality of drinking water within them, therefore it is essential that firefighters have a thorough understanding and appropriate skill level.

Once a water supply has been established, firefighters need to then learn the skills to transport the water to the fire, to use the right technique and means of applying the water and the use of pumps to give the water momentum and pressure.

The first part of this process will be to get the water to a pump. In the case of open water, this will be through a large-diameter hose that is capable of withstanding a negative internal pressure (vacuum). This type of hose is known as a suction hose, and the end to be placed in the water source will be fitted with some means of filtering solid objects. The outlet end will attach directly to a pump inlet. Firefighters will learn how to attach the hose to the pump, how to effectively locate the inlet within the water source and how to tie the hose in such a way as to ensure it is not placing a strain on the pumps or connections when full of water. Firefighters will also use the lines to secure and take the weight of the hose to adjust or remove the submerged inlet.

Where the water supply is from a hydrant, firefighters need to know how to use flat laid delivery hose to connect to the pump, where to connect it and how to control the pressure through the pump.

Water entering the inlet side of a pump is identified as the supply. Once it has been through the pump, it is energised and is controlled through a valve to supply firefighters, which is known as the delivery.

When firefighters lay delivery hose, they need to consider many factors that will include avoiding potential obstructions to the water supply. They also need to consider the impact of height rise on pressure. It must be laid in such a way that it can be manoeuvred as progress is made through the fire attack.

Fire hose itself will, of course, deliver water from its outlet. However, in order to give the water the appropriate throw, jet stream and body, firefighters attach an outlet known as a branch to the end of the hose. The branch provides a valve to control the flow of water through the hose line and is able to create either a solid jet with the power and range to travel significant distances or a misting pattern that increases the surface area of water droplets and therefore the water's ability to absorb heat.

Firefighters learn the appropriate branch needed for different types of fire attack, they will learn how to overcome changes in height through the use of intermediate pumps and how to increase flow through wider-diameter hoses or increasing the number of hose lines available.

Ladders and working at height

The next set of basic skills firefighters will acquire is in the handling of manual ladders. Depending on the type of fire appliance in use, ladder lengths will range from around 9m (30ft) up to 13.5m (44ft).

Firefighters will learn how to safely and swiftly remove ladders from the fire appliance, how to safely lift and carry ladders and how to elevate then raise them before pitching them against a solid structure.

Once firefighters have mastered the correct operation and use of ladders they will then incorporate ascending heights, operating in areas that are difficult to reach, moving from ladders to windows and over walls, as well as learning how to use ladders for the rescue of people from height.

Combining practical skills

From basic skills in the use of fire hydrants and running out hose, trainee firefighters will move on to use ladders and pumps before combining these skills to the point where they are able to get a full range of firefighting equipment to work, to resolve issues that may arise and to achieve objectives set as part of exercises that combine the different basic skills.

Completion of an initial skill is known as acquisition and after this, firefighters will go through a development programme where they practise and demonstrate a new skill. Once they have consistently demonstrated the skill through simulated and workplace assessment, they will be deemed competent. From this point onwards, to remain competent the firefighter will be required to maintain these skills and this forms part of ongoing continued assessment. This ensures that skills are not allowed to fade, particularly for activities that may occur infrequently but require instant action.

LEFT Firefighters pitching a 13.5m ladder. *(Devon and Somerset FRS, Minehead)*

Theoretical knowledge

In addition to gaining practical skills and knowledge in the safe use of techniques and equipment, it is essential that firefighters understand the basis of a range of mathematical, scientific and engineering principles to underpin their knowledge of fire behaviour, equipment use, electrical hazards and building construction.

BELOW Firefighters combining their practical skills using multiple ladders, hose lines and branches. *(Alfie Martin)*

ABOVE Theoretical training is undertaken in the classroom environment. *(Devon and Somerset FRS, Minehead)*

Theoretical knowledge may be learned through accessing written and electronic information, through the use of multimedia and through practical vocational scenarios. For example, in order for a firefighter to understand the physical effects and impact of passing a water supply through lengthy hose lines, they need to understand the principles of friction and friction loss. In addition, to understand why a branch can become dangerous and difficult to control under pressure, there is a need to understand jet reaction. The underpinning theory of basic physics is fundamental to understanding practical challenges that have to be addressed in firefighting.

In relation to ladders, there is a need to understand angles in order to ensure ladders are positioned safely, and to understand basic physics in order to assess safe loading and use of ladders at a range of lengths and angles.

THE SCIENCE OF FIRE IN SIMPLE TERMS

Many of us will have learned about the triangle of fire during our early education. This model demonstrates that in order for combustion to occur there needs to be three elements present (the three sides of the triangle). These are fuel, heat and oxygen. The triangle of fire also shows us how each of the three elements work together in creating and developing a fire. It teaches us that removing any one of the sides of the triangle will prevent or extinguish a fire. As a reminder, here is how the triangle comes together for combustion to occur:

RIGHT Diagram of the fire triangle *(Shutterstock)*

Fuel

For a fire to start there must be something to burn – this is referred to as the fuel. Fuel has to be a combustible material and can be in solid, liquid or gas form. Solids include wood, paper, fabrics, plastics and rubber. Liquids may be oil, petrol, diesel, spirits or derivatives of these such as paints, glues and industrial chemicals. Gases may be in the form of liquefied petroleum gas, natural gas, aerosol propellants and other flammable gases.

Heat

Where fuel is present, heat is also required for a fire to start. All combustible materials release flammable vapours when heat is applied; it is these vapours that ignite and burn. Heat is also responsible for fire to spread and for it to continue to burn by taking moisture from nearby fuel and warming fuel that is nearby. As the temperature of a fire increases, the speed of fire spreads and intensity will increase also.

Oxygen

Along with fuel and heat, a fire also requires oxygen to burn. The air we breathe is made up of approximately 21% oxygen. For a fire to burn it needs around 16% oxygen.

and ongoing workplace assessment. Many fire stations have the facilities for basic training, both practical and theoretical. In addition, firefighters will train using venues in their locality. This enables firefighters to practice in a range of conditions and supports the development of local knowledge.

The development and maintenance of theoretical knowledge is delivered through information based on sources including national guidance, technical documents or local risk. This type of training is often delivered through computer-based e-learning programmes.

Fire and Rescue Service activities

The incidence of fires, fire deaths and other fire-related calls have seen an overall decline over the past 20 years.

Changes in building design, building regulations and furniture safety have made the majority of modern buildings safer. In addition, smoke alarm provision and ownership has dramatically increased life safety in the home.

The advances in commercial fire detection and alarm systems in public and commercial buildings have also reduced the numbers of false alarms while increasing the capability to detect and locate genuine fires in their early stages.

The modern fire and rescue service has adapted from being a primarily reactive emergency service to a position where firefighters are now very proactive in addressing the risks associated with fires and other emergencies. In addition, data analysis using historic and predicted information allows services to target activity towards people and buildings that represent the greatest risk.

This activity is generally managed through a programme of prevention and protection activities.

Prevention

Prevention of incidents can be achieved by a number of methods. Technology is an important factor and is primarily aimed at changing human behaviours. Prevention focuses on raising awareness of risk and providing appropriate advice and information to protect people and property from a wide range of activities and incidents that would otherwise cause harm.

Prevention activities are designed to help protect people from birth to the end of their natural life and successful prevention work carried out by the fire and rescue service alongside partners is credited with reducing injuries, deaths and destruction of property.

Prevention activity can be broadly described in the following categories:

- Education
- Intervention and assistance

ABOVE The delivery of fire safety education is undertaken by specialist officers as well as response crews. *(Cornwall FRS)*

Education

Education used as a preventative tool involves the providing of information and skills that will enable people to learn about hazards and risks. This knowledge will then help people to recognise and understand dangerous situations and activities and how to reduce or avoid them altogether. Prevention education activities will include:

Early Years Programmes

Programmes aimed at young people just starting their lives at school will focus on simple life skills such as not playing with matches, lighters or fire, what to do if there is a fire in the home and what the fire and rescue service does. These programmes may be delivered by the local fire and rescue service, by teachers in schools or by leaders of youth groups.

In addition, fire and rescue services have produced books and videos that share safety

RIGHT A firefighter of the future – happy to be learning about fire safety at school. *(Louise Collins, Dunster First School)*

messages within simple stories using characters to make learning enjoyable and easy to access.

Education through school years

As children evolve through their school years, age appropriate messaging continues to help young people to learn and develop their life skills for the dangers and temptations they may face. This will include understanding what to do in the event of an emergency such as an escape plan for their home, how to test a smoke alarm, understanding the dangers and consequences of deliberate fire-setting, and other dangers.

Further education

During further education, young adults are provided with messaging around road safety as they work towards and gain their driving licences, as well as the importance of their own safety as passengers. There are some incredibly innovative and immersive programmes delivered to this group, often in conjunction with other emergency services.

Adult education

As people establish their own homes and families, messages evolve to help them live safely. Often this will be a reinforcement of previous programmes of education with a focus on specific dangers such as cooking, gas and electrical safety, chimney fires and carbon monoxide dangers.

Specialist campaigns are used to highlight specific messages. For example, how to correctly install a child car seat or how to safely deal with a fire in a chip pan.

In order to ensure thorough and fair access to information, fire and rescue services will use specialist messaging and delivery techniques to those who might otherwise find it difficult to access. Wherever possible, information will be delivered in a range of languages, using audio for those with impaired vision, and other assistive technology.

Intervention and assistance

Prevention intervention and assistance activities involve firefighters engaging with specific 'at risk' individuals, families or groups and working with them to reduce that risk.

Visiting someone in their home to carry out a home safety visit has proven to be an effective way to reduce both risk and incidence of dwelling fires. Such visits will generally be carried out by firefighters or other uniformed staff who will carry out a visual check of the home and identify potential hazards. Working with residents, these visits will highlight potentially dangerous circumstances such as blocked escape routes, dangerous storage or unsafe electrical devices. In addition, a home safety visit will identify opportunities to improve safety in the home, which may mean simple changes to the location of items such as a toaster through advice to the residents, and with the installation and testing of smoke alarms. Many fire and rescue services also work with care and health providers to complete simple assessments of individuals that could lead to further support from other agencies.

BELOW Fire and rescue services complete thousands of home safety visits each year. *(Cornwall FRS)*

LEFT Working with young people to reduce fire-setting behaviour has proven to be successful. *(Shutterstock)*

Another example of intervention is working with young people who may develop an unhealthy interest in deliberate fire-setting. Firefighters and other staff are trained to work with young people to explore and challenge behaviours that could lead to deliberate fire-setting and the consequences of such behaviour.

Fire and rescue services aim to deliver safety programmes, interventions and assistance across the whole of society from the youngest to those in later years.

Protection

Fire and rescue authorities have powers invested to enforce legislation under the Regulatory Reform (Fire Safety) Order 2004. Fire protection is a term used to describe the activities that fire and rescue services undertake to ensure businesses comply with the legislation, which is designed to ensure that buildings, structures and venues covered under the legislation are safe for people to live in, work in or visit. Notably, the Regulatory Reform (Fire Safety) Order 2004 does not apply to single private dwellings, although it does cover shared access routes where a property has more than one individual dwelling.

Meeting the requirements of the legislation is the responsibility of the 'responsible person' for a building, who must ensure that measures are taken to reduce the risk of fire and the risk of spread of fire, measures in relation to means of escape from the premises, including safe escape routes and emergency escape lighting, measures in relation to fighting fire and measures in relation to the detection of fire and the ability to warn of fire. In order to assist with this, fire and rescue services may carry out a range of activities to support compliance with the legislation. Guidance is made available, generally online and free of charge, while compliance events may be held for different types of premises to provide advice and improve understanding of those with a responsibility for the building.

Where there is a failure to comply with the legislation, fire and rescue authorities are empowered to act to enforce the legislation and to prosecute offenders. A warranted enforcement officer may prohibit the use of a building or part thereof, may take samples and may gather evidence where a breach of the legislation is suspected.

BELOW To the untrained eye, fire protection measures are not always as obvious as this extinguisher. *(Philip Martin)*

Fire precautions

In order to meet the requirements of the Regulatory Reform (Fire Safety) Order 2004 there is a wealth of standards, guidance and technology that can make our buildings safer places.

Buildings are planned, designed and built in accordance with relevant standards and regulations. The formal process of risk assessment is a continuous responsibility of the appropriate responsible person when a building is in use. The fire risk assessment will identify fire hazards within and associated with the building. Where a risk is identified the risk assessment process will be used to reduce the risk to the lowest level deemed reasonably practicable. In achieving this the appropriate control measures will determine what active and passive fire precautions measures can be put in place as well as any management actions to ensure the safety of people in the building if a fire should start.

Passive fire protection

Passive fire protection exists in the form of fire-resisting walls, fire-resisting doors, seals on doors, specially insulated wiring and other non-electrical or mechanical building features. Passive fire protection is designed to create barriers to the spread of fire and smoke in the event of a fire.

RIGHT Fire door and frame fitted with an intumescent strip and smoke seals. *(Alfie Martin)*

Fire resistance

The term fire resistance describes the ability of a structure or material to withstand exposure to a fire or to provide protection against it.

The fire-resistant properties of a structure or component such as a door, floor, ceiling or window will be described as its fire resistance rating. This rating is usually expressed in hours.

Fire-resisting walls

Fire-resisting walls are required where it is necessary to create a barrier to prevent the spread of fire and/or to maintain integrity if exposed to fire. Fire resistance may be achieved using a range of building materials that are both natural and man-made.

Fire exit routes from a building, including staircases and other enclosures, will be required to achieve fire resistance. In addition, in order to minimise the spread of a fire, the building will be sub-divided into compartments. These compartments will be separated by fire-resisting walls.

Fire-resisting doors and other openings

Where it is necessary to breach a wall or compartment to provide access and egress, it is critical that the integrity of the wall is matched by the door, glazing, ducting or services (such as cables, pipework or ducting) that closes the breach.

In the case of fire doors, both the integrity of the door and the frame in which it is held are critical to maintaining resistance in the event of a fire. A gap between the door and frame is required for operation, but in the event of a fire this could allow the passage of smoke and heat leading to an early compromise of the fire resistance. To protect the doorway, fire doors or their frames have an intumescent strip fitted into a recess around the door or frame. The material used in the intumescent strip will expand when exposed to heat, closing the gap between the frame and the door.

Where services carried through cables and pipework have to pass through a fire-resisting wall, both these and the opening through which they travel must be capable of maintaining the fire resistance. This is achieved by the use of cables and pipes manufactured using fire-resistant materials and through the use of

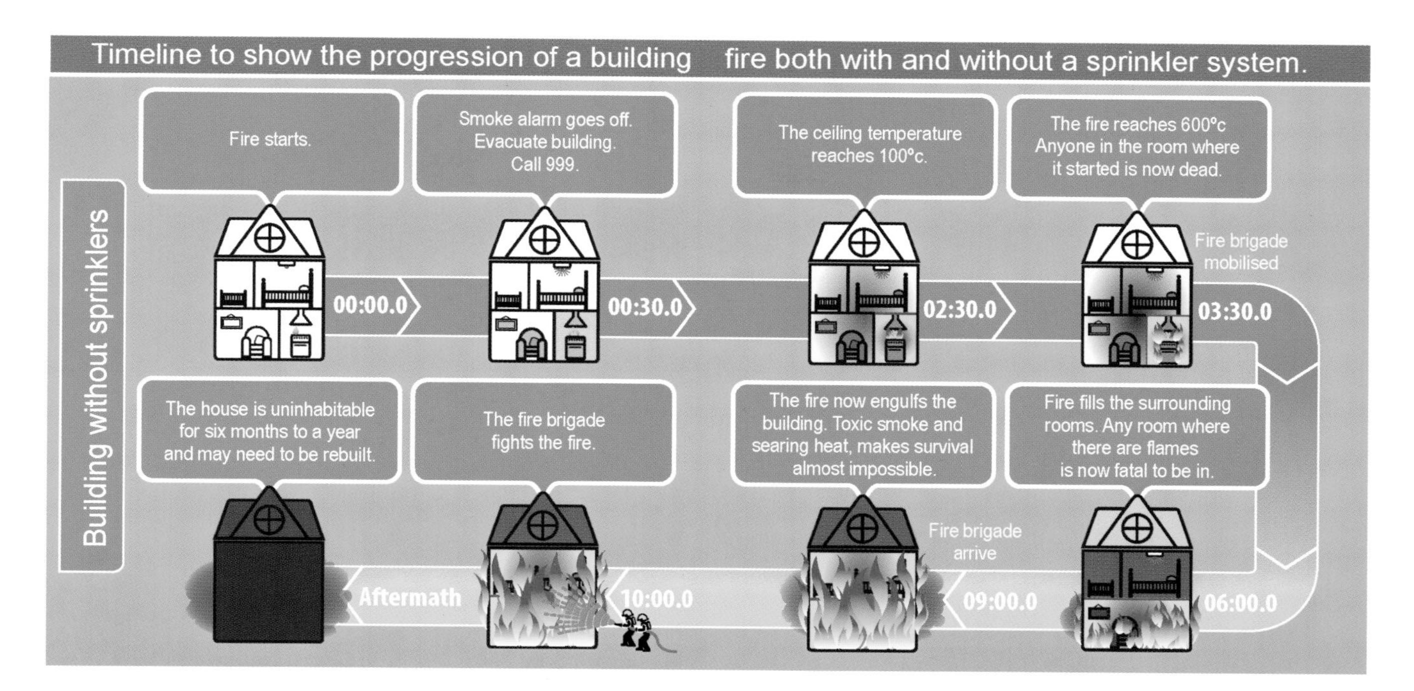

unlike the image we are too often exposed to in the movies.

Another less common type of activating mechanism is the use of a fusible link that will melt in heat. This performs the same function as a bulb and the system is therefore very similar in all other aspects.

Sprinklers have proven to be extremely reliable over their long history. More than 70 million sprinklers a year are installed worldwide, although the UK accounts for less than two million per year. Research extending over 100 years demonstrated the reliability of individual sprinklers to be a commendable 99.7%.

In addition, fires in a building where sprinklers are fitted and operate are most often extinguished by the operation of a single sprinkler head with damage limited to a minimal area.

In the UK, where sprinklers remain uncommon compared to the rest of the world, sprinklers have been proven to reduce property damage by around 90%. In addition, and more importantly,

ABOVE AND BELOW Illustrated timeline for buildings with and without a sprinkler system. *(Devon and Somerset FRS)*

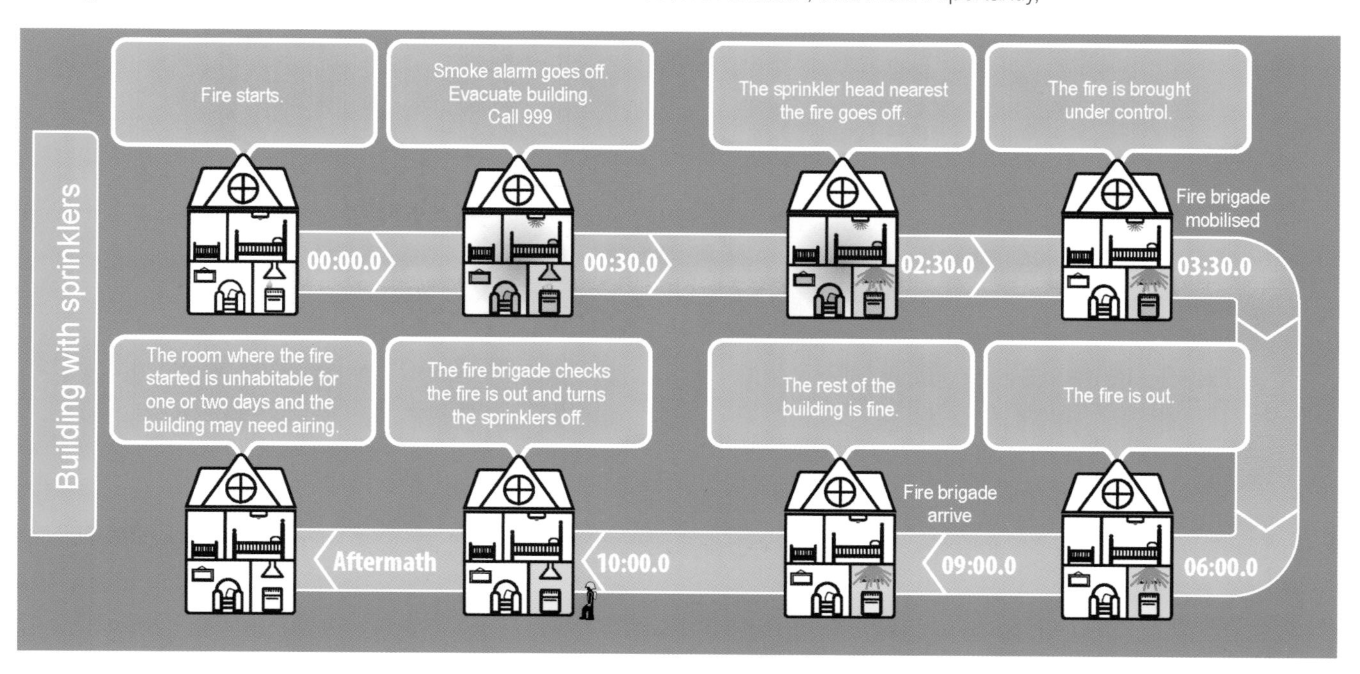

ABOVE A fire drencher system operating at a building under construction – the mixing of the smoke and water creates a mist. *(Shutterstock)*

sprinklers can reduce injuries by at least 80% and sprinklers reduce the risk to firefighters.

Smaller fires and minimal use of water delivers an additional benefit in the form of reduced environmental damage.

Other automatic extinguishing systems

Other types of extinguishing systems are in use for different types of risk and these include:

Drencher systems

A drencher system is used to protect the outside of a building from fire spread. A drencher is a system of outlets positioned to protect roofs and external openings such as doors and windows.

Water spray systems

Using a system of pipework and water spray outlets, a water spray system can provide protection to outdoor plant and equipment such as transformers. These will contain and potentially extinguish a fire.

Water mist systems

Water mist systems create small water droplets to control and extinguish fires. In operation, a water mist system will both cool and create a barrier between the unburned fuel and radiant heat. Due to the additional pressure and reduced droplet size in water mist, these systems will use less water compared to traditional sprinkler systems.

Inert gas systems

An inert gas is a gas that has extremely low reactivity with other substances. This makes it suitable for extinguishing fires in a contained

RIGHT Precision-engineered nozzles are used with water-mist and inert-gas systems. *(Shutterstock)*

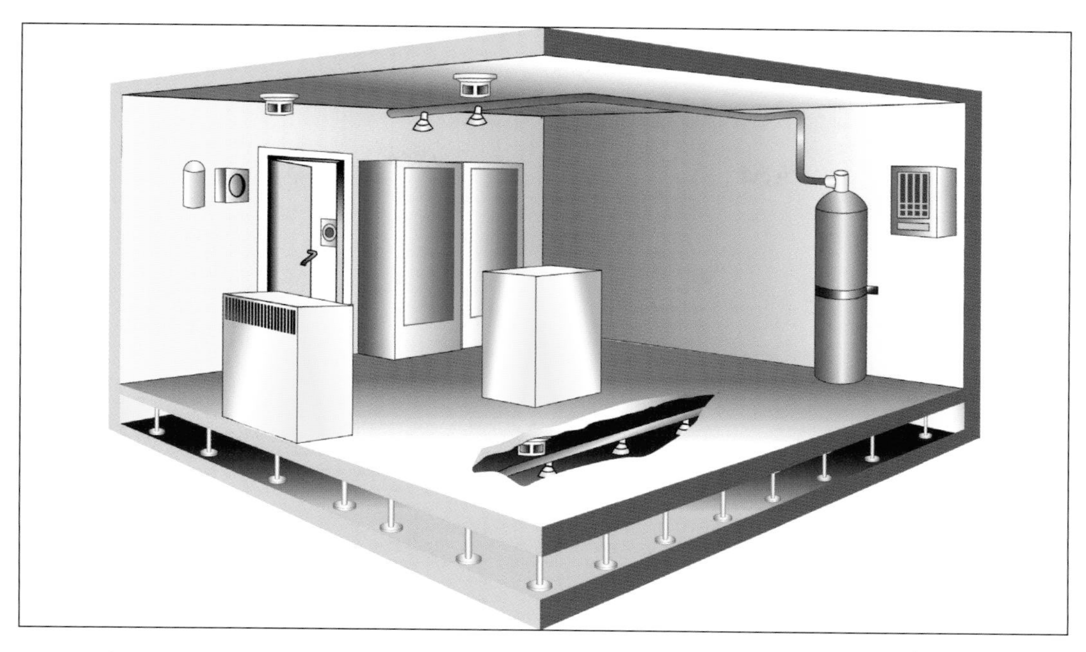

LEFT **Illustration of a gas suppression system protecting a computer server room.** *(3M)*

area such as a computer server room, an area with electrical components, stored gases or other sensitive equipment.

The activation of an inert gas system will reduce the oxygen within a sealed room to below normal air levels (less than 15%). This leads to insufficient levels of oxygen to sustain class A, B and C fires. At this reduced level, while a fire cannot burn, it does remain possible to breathe should someone be in the area at the time of a system operating.

Chemical systems

Specific chemicals are able to extinguish fires by interfering with the physical combustion processes. The molecules in the chemical used will absorb heat to reduce the temperature of the flames and extinguish the fire.

Domestic fire suppression systems

Using the same principles as the commercial suppression systems described above, the installation of sprinklers and other suppression systems in dwellings is easily achievable. The installation of such systems in dwellings has been commonplace in the USA for many years and there are some examples of developments within the UK that have adopted sprinklers and other suppression systems as a standard, as well as local authorities and housing associations that install systems within their property portfolios.

Currently, only Wales has a requirement for automatic fire suppression systems to be installed as part of the principality's building regulations.

While the installation of sprinklers is most efficiently achieved during the construction

BELOW **Illustration of a domestic sprinkler head and its operation.** *(Domestic Sprinklers Cornwall Ltd)*

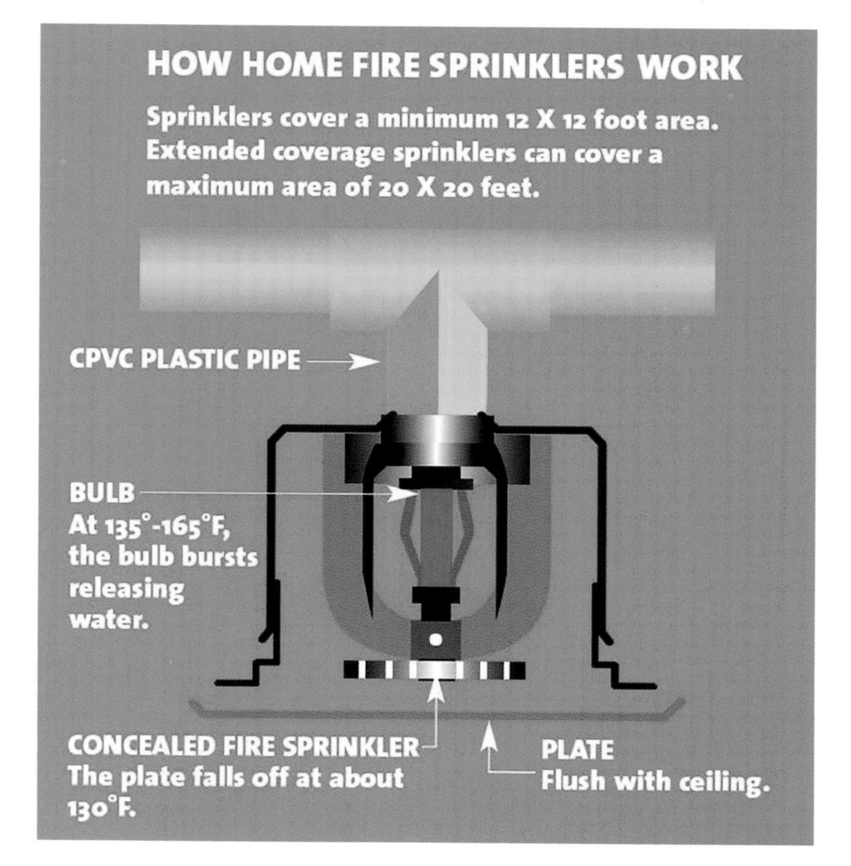

phase of a property, it has been proven to be both feasible and not prohibitively expensive to retrofit systems in existing buildings.

Domestic fire suppression systems use components that are designed to be cosmetically unobtrusive and some systems would not be seen by the untrained eye.

Response

There has been a great deal of success in improving fire safety and in reducing the incidence of fires as a result of the effective prevention and protection work. However, when there is some kind of failure, whether through human or mechanical failure, or where there is a disaster that results in injury or entrapment, the fire and rescue service will generally be expected to respond.

Emergency response probably remains the most well-known role of the fire and rescue service. Daily headlines often tell the story of firefighters being called to an incident at a local, national and sometimes international level. Firefighters have evolved over many years and the emergence of new technology has helped them to deal with ever-more complex scenarios and incidents well beyond the scope of fires.

RIGHT The need to attend incidents quickly must be balanced with the safety of the crew and other road users. *(Shutterstock)*

by all services in developing effective response plans.

Joint Decision Model

The Joint Decision Model was developed to support commanders and to create a common understanding of an incident. While this model is used by all agencies, there will be different processes by which each service may gather information to support their decisions. Each service will complete a log, which will record the decisions made by their respective commanders. Where joint decisions are made, these will be recorded in a 'joint decision log'.

Joint understanding of risk

The understanding and reduction of risk is key to the resolution of any incident. A shared and acknowledged understanding of incident risk is achieved through shared information.

Commanders must be aware of the range of risks and hazards that each organisation has identified and is dealing with. Having a joint understanding of risks and hazards supports prioritisation of activity and the implementation of appropriate control measures to mitigate risk. This creates a safer working environment for all personnel at any incident.

Shared situational awareness

Through good communication and coordination, which is supported by using both METHANE and the Joint Decision Model, all responding agencies are able to see a common picture that has been developed and agreed by working together. With this information, the services are able to prepare an agreed response plan that has taken account of all known threats and hazards. In addition, it will have been created with the knowledge of what resources and capabilities are available and what decisions and actions can be lawfully undertaken to achieve resolution.

Working with other agencies

Often an emergency incident will require specialist knowledge to be resolved. While the fire and rescue services have the knowledge, skills and procedures to deal with the initial emergency response, the incident resolution may require the specialist knowledge and skills of other professionals to assert a final resolution.

Examples of this will include incidents involving the transport network, chemicals and mains services, such as gas and electricity and specialist processes.

LEFT Police, ambulance and fire command vehicles will be located together at major incidents to ensure that effective communication and decision making is achieved. *(Philip Martin)*

CORNWALL
FIRE & RESCUE SERVICE
Community Fire Station
Gorsav Tan Kemenieth Talvadn
Tolvaddon

Chapter Four

The fire station

Wherever we are in the UK, we will be protected by crews primarily based in our nearest fire station. The fire and rescue service's integrated risk management plan will have determined, affirmed or modified the location of any fire station, which will be based on risk and response requirements. For those living in larger towns and cities the fire station may be just a short distance away, while in rural areas they may be further away.

OPPOSITE Modern fire stations are designed to meet the needs of the whole community.
(Cornwall FRS)

RIGHT **Modern fire and rescue service headquarters are designed to meet the needs of the diverse workforce and the environment. Solar panels can be seen on the roofs.** *(Cornwall FRS)*

BELOW **A typical appliance room.** *(Cornwall FRS/Kate Shorey)*

Crewing arrangements and other factors will determine the size and functions of a fire station, and increasingly these have become shared facilities with multiple uses.

As a basic model, the fire station serves as a local base in which fire appliances can be securely kept and maintained, as well as a base for providing welfare and storage facilities for the firefighters who work there. Most fire stations will also have an outside area for routine training in the use of pumps and ladders as well as an indoor room for theoretical training.

In order to provide an understanding of the available facilities, we will look at a modern fire station that is crewed by both wholetime and on-call crews. We will look at the internal and external facilities and their use.

Internal facilities in the fire station

Appliance room

The appliance room is a firefighter's most familiar part of a fire station. This acts as the garage for fire appliances and will have good access to the road network.

In order to ensure fire appliances are ready for immediate use they remain plugged in to mains electricity when parked in the fire station. Mains power is used to supply a trickle charging facility for the vehicle batteries, which are likely to be connected to charging facilities for specialist equipment.

As with most large goods vehicles, fire

appliances are fitted with air powered brakes. These rely on a pressurised system to operate and a lack of pressure will leave brakes in a fail-safe position. In order to ensure braking systems are constantly operable, fire appliances have a built-in mains-powered compressor that maintains a consistent minimum pressure.

Recognising that fire appliances may be required to go from being inactive to being used with peak power demand in a very short time, there is a need to keep engine oil viscous and ready for high demand use from start-up. To achieve this, some fire appliances will have heating elements in the engine sump, maintaining oil at a minimum temperature.

As well as being a garage, the appliance room provides a useful, large indoor space. This is often used for practical training activity, with some fire stations designed for use in specialist training.

ABOVE Fire stations have meeting rooms which are available for use by community groups. *(Alfie Martin)*

Station office

Most fire stations will have an office that acts as the base for station activity. This room will generally provide access to the fire and rescue service ICT network, acting as the general hub for information exchange, administrative activity, station activities, training records, online training and incident reports.

Activity reporting and recording is critical in ensuring performance standards are recorded and maintained. Whether at the completion of training, a prevention, protection or other risk-reduction activity, recorded information is valuable for use in monitoring performance and in directing future activity.

A practical example of the need and purpose for recording relates to a change in risk. When the use of a premises changes in a way that affects risk, the information about that change may be critically important in the event of a fire or other incident. A change in risk may necessitate a change in response to an incident. There may be additional training requirements for firefighters and fire precautions may need to undergo review. Having access to this information ensures that changing risk is both understood and mitigated.

Training/meeting room

Theoretical training, presentations, meetings and other classroom-based activities are delivered in a training room in the fire station. Training rooms will generally be fitted with audio/visual presentation facilities and desks.

Training rooms are often used for presentations to community groups and other organisations and, where practical, these

BELOW When alerted to an incident, crews head for the muster bay where their kit is stored. The pole drop provides a quick descent from the upper floors. *(Cornwall FRS/Kate Shorey)*

RIGHT Entrance to the pole drop. *(Alfie Martin)*

rooms may be made available for use by other organisations to optimise their use and maintain engagement with the local community.

Muster bay

The description 'muster bay' has its origins in the naval heritage of the fire and rescue service. The term 'muster' was used to describe the gathering of sailors or other military personnel, and the muster bay in a fire station is where crews keep their personal protective equipment and therefore muster at the start and end of their duty.

Modern muster bays provide the facilities for firefighters to store their firefighting gear in a way that makes it easy to access and promotes continuous airing. This room will usually be immediately adjacent but separate to the appliance room, supporting a swift response. In the case of a wholetime fire station, this may be the area that is accessed by a pole from upper floors.

The effects of smoke and other products of combustion are recognised as being potentially harmful to firefighters where firefighting gear has become contaminated. For this reason, firefighters need to ensure potentially contaminated firefighting kit is properly cleaned after use. Some fire stations will have the facilities to carry out cleaning on site. Fire stations have drying rooms to ensure firefighting kit can be dried quickly and ready for use as soon as possible.

The proximity of the muster bay to the fire appliances often makes this the ideal location

ABOVE Fire station kitchen and dining room. *(Alfie Martin)*

RIGHT Fire station fitness room. *(Alfie Martin)*

for mobilising messages to be received by the station printer.

Pole drop

The firefighter's pole is still a feature at many multistorey fire stations across the globe. The pole provides a swift route between levels and may be staged depending on the number of floors to be descended.

Fitness room

Most fire stations will have some basic fitness equipment to support ongoing firefighter fitness. Where possible, a fire station may have a fitness room housing a range of strength-building and aerobic fitness equipment.

Kitchen dining area

Many fire stations will have basic food preparation and cooking areas for firefighters working shifts. A particular feature in most fire stations will be the provision of an auto-cut-off device that will isolate cooking appliances in the event of the station call-out system operating.

Study rooms

Wholetime fire stations may have study rooms that can be used for individual research, online training and administrative work. They may also be used as a place to relax during allocated rest periods.

ABOVE Fire station study and rest room. *(Alfie Martin)*

External facilities in the fire station

Drill yard

Where building space allows, the outside grounds at a fire station will have been designed for use as a training area. It is common to see a training tower with stages and windows to enable firefighters to practise their ladder and firefighting skills, training in the use of safety at height equipment and practising rescues from height.

The presence of a fire hydrant allows for training in the use of pumps and hose lines as well as providing a convenient location to maintain the fire appliance water supply.

Recognising the wider role, drill yards provide a safe location for training in vehicle extrication, dealing with hazardous materials, setting up technical equipment and the use of a range of specialist appliances.

With planning and imagination, fire station drill yards can be configured to represent various fire and rescue scenarios, which enable crews to practise the combined use of a wide range of equipment.

Forecourt and road access

Maintaining road access from the fire station is

LEFT Firefighters undertake combined ladder and hose drills using the training tower. The tower's multi-level platforms allow for more complex training scenarios to be achieved. *(Cornwall FRS, Wadebridge)*

ABOVE Warning lights to stop traffic are provided outside fire stations located on busy roads, to ensure firefighters can respond quickly. *(Steve Greenaway)*

critical to an effective emergency response. Where fire stations exit onto busy thoroughfares, it is common to see red flashing lights that will stop traffic to allow safe egress. In addition, there will often be a yellow hatched area to the front of the fire station in order to maintain a clear exit route.

Combined emergency service stations

The design requirements and location for emergency response are often similar across the blue light services. Recognising the potential for more efficient use of estates, there are increasing numbers of emergency services who are sharing facilities. In some cases, this has been achieved using existing facilities with minor modifications to meet specific needs. Where services are developing new sites they are engaging with other blue light services and where appropriate this may lead to a co-designed facility.

Experience of these arrangements has demonstrated that beyond the economic efficiencies, co-location supports joint working and response.

Community use of fire stations

Having recognised the potential to increase the use of fire stations for public use when not being utilised for their core function, many fire and rescue services have invested in upgrading fire stations to make them suitable for use by local communities.

Where practicalities such as security and welfare provision can be addressed, the wider use of a fire station provides a great facility for supporting communities and community groups.

A number of fire and rescue services have incorporated wider community use into the design of new buildings. This has proven very successful, providing valuable shared facilities to bring communities together and in contact with their local FRS.

TRI-SERVICE SAFETY OFFICERS

Through the integrated risk management planning process, the town of Hayle, in Cornwall, was identified as falling outside of an acceptable response area for Cornwall Fire and Rescue Service. Following discussions with police and ambulance a brand-new combined facility was co-designed and built to house all three services.

Seeing an opportunity to further build on a shared physical resource, all three emergency services committed to a pilot evaluation of a tri-service safety officer (TSSO) who could primarily deliver preventative activity for the Hayle community but also provide a shared response capability. The appointed TSSO was already a trained on-call firefighter and went on to complete training in elements (but not all related functions) of a police community support officer (PCSO) and training as a community responder for the ambulance service.

LEFT A tri-service safety officer and response vehicle. Note the use of the blue, green and red livery on the vehicle to depict all three services. *(Alfie Martin)*

The trial underwent significant scrutiny throughout and proved to be a successful model with benefits exceeding those originally envisaged. The role has, to date, been replicated across a further ten communities in Cornwall.

BELOW Hayle tri-service station in Cornwall – the first purpose-built station for fire, police and ambulance. *(Cornwall FRS)*

EMERG

Chapter Five

Equipment

As the range of incidents attended by the fire and rescue service has increased and diversified over many decades, so has the specialist equipment that is available for use.

OPPOSITE Large and complex incidents often require the attendance of multiple appliances. *(Shutterstock)*

ABOVE Modern fire appliances carry a vast range of equipment, as well as specialist crews, to deal with many varied incidents – as demonstrated by the personnel from this two-appliance station. *(Devon and Somerset FRS, Minehead)*

Early firefighters relied on very basic manual tools to create firebreaks before the evolution of basic pumps and beyond. Modern firefighters have the capability and equipment to respond to and deal with the widest range of emergency incidents.

Personal protective equipment

In order to be safe at an incident and to comply with health and safety regulations firefighters must have appropriate personal protective equipment (PPE).

This will consist of protective footwear, trousers and jacket, gloves for hand protection and a specialist balaclava and helmet with built-in safety spectacles to protect the head and eyes. Additional protection in the form of self-contained breathing apparatus is required if working in contaminated and irrespirable atmospheres, and gas-tight suits need to be worn when dealing with certain chemicals in a range of forms.

Compartment firefighting kit

The last three decades have seen revolutionary changes in the protective clothing issued to firefighters for dealing with fires in buildings. Modern firefighting kit is made from flame-retardant material with lightweight insulation, constructed to be breathable and water resistant while providing a high degree of thermal protection.

Modern firefighting kit is made in a wide range of sizes, ensuring it is well fitted to the wearer regardless of their individual physical characteristics. Additional safety is afforded to firefighters through the addition of padding and protection for the knees and elbows,

ABOVE For all fire-related incidents, firefighters will wear a uniform know as Compartment Firefighting Kit, which is designed to provide enhanced levels of thermal protection. This uniform can be supplied in differing colours and fabrics, although the levels of protection provided will all meet the required standards. *(Bristol Uniforms)*

protective cuffs to protect forearms and high conspicuity markings.

Firefighting boots provide protection to the toes and soles through the use of reinforced materials while styles and designs provide the best flexibility for the wearer. Thermal protection, waterproofing and wearer comfort are paramount, along with durability.

Protection to the head and neck is provided in two ways. The first level of protection is a flame-retardant fire hood that is worn like a balaclava. This is worn with the neck section tucked into the collar of the firefighter's tunic to ensure that no skin is exposed.

Impact and eye protection are provided by a fire helmet with integrated safety glasses and a full-face visor. The helmet is fully adjustable to the individual wearer and provides protection from impact to the wearer's head.

Firefighting gloves are designed to provide both physical and thermal protection. They need to maintain good dexterity in order to perform certain tasks, while wearing gloves requires a high degree of flexibility.

Technical rescue kit

For many years, most fire and rescue services have used compartment firefighting kit as the standard protection for firefighters in all situations. While this kit provides protection from heat, flames and water, the thermal protection can restrict some movement and become uncomfortable in tasks where the wearer is completing an arduous activity or operating in a confined space.

Many services are now adopting specialist rescue kit that has been designed to provide the required level of protection in situations where thermal protection against the heat of fire is not necessary, but hazards exist in the

RIGHT Technical rescue uniform is designed to provide the firefighter with a greater level of mobility without the need for thermal protection. *(Bristol Uniforms)*

form of sharp edges, contaminated fluids and blood products. The absence of a requirement for thermal protection enables the garments to be designed and manufactured in a way that makes the kit lighter to wear, more agile and therefore suited to more technical rescue environments.

Water rescue kit

Entry to water requires firefighters to have appropriate PPE for protection from extreme cold, which could otherwise cause cold water shock, hypothermia or extreme fatigue.

Water rescue kit consists of a layered approach, incorporating a base layer of a thermal undersuit and socks. This layer ensures the wearer remains warm. The outer layer consists of a specialist drysuit with integral boots and water-sealing cuffs to the wrists and neck.

Hands are protected with water rescue gloves and the head by a water rescue helmet. The helmet has to be of a specific design to ensure the wearer's head is adequately protected without affecting the buoyancy of the wearer or enhancing the risk of neck injury.

A personal flotation device (PFD) is worn, incorporating a rescue harness to which a safety line can be attached. The PFD has concealed flotation pockets that can be automatically inflated through a disposable compressed gas container fitted within the device.

Gas-tight suits

Gas-tight suits (GTS) are provided to protect firefighters from exposure to potentially poisonous substances occurring as a result of an uncontrolled incident, or in the event of a malicious terrorist chemical attack.

The GTS is designed as a one-piece coverall with integrated gloves and foot protection that enables the use of separate boots. A GTS is worn in addition to standard PPE appropriate

BELOW Firefighters wearing water rescue suits rescue a casualty from a river using a throw line. *(Cornwall FRS)*

BELOW RIGHT Gas-tight suits provide firefighters with additional levels of protection when dealing with incidents involving hazardous materials. *(Devon and Somerset FRS, Minehead)*

for the task, along with breathing apparatus, all of which are contained within the suit.

Gas-tight suits afford a level of protection at incidents involving the leak of a dangerous substance during use, transport or storage.

Fire and rescue helmets

Firefighting helmets have been iconic symbols of the firefighter for many years. The shape depicted in Roman times evolved slowly with similarities remaining until the 21st century. Over more recent years the design and functionality has developed to create a headpiece that provides comfort to the wearer, alongside optimum protection with the means to integrate with other personal protection.

The modern fire and rescue helmet provides full head and face protection through an outer shell made of strong and durable composite materials. It includes both a full-face polycarbonate visor and eye protection in the form of spectacles, either of which can be used independently. The fully adjustable internal harness can be adjusted by the wearer with one hand. Neck protection is provided through an insulated flame-retardant fabric collar. In addition, the helmet has a torch incorporated into its design and can be integrated with a breathing apparatus facemask.

Protective gloves

Hand protection is critical in fire and rescue operations. Firefighters will have at least two pairs of gloves; one specifically for firefighting operations, which afford the wearer both thermal and sharp end protection, and a more generic protective glove for other operations.

ABOVE Modern firefighting helmets resemble a motorcycle helmet, and are designed to be light weight and extremely strong. *(Devon and Somerset FRS, Minehead)*

Firefighter boots

Firefighter boots have become another specialised piece of personal protection for the wearer. In addition to the need to be waterproof, the nature of fire and rescue operations creates a need for additional protection in the form of reinforced soles to prevent the penetration of sharp objects, protected toecaps in case of falling objects, resistance to chemicals and hydrocarbons, and sufficient flexibility to maintain wearer comfort during extended periods of use.

Breathing apparatus

Breathing apparatus represents one of the most important items of personal protection for firefighters. With some fires producing chemicals such as hydrogen cyanide and other noxious gases, the risk to firefighters from even small amounts of smoke inhalation is very high. In addition, fire gases reach extremely high temperatures, making breathing impossible.

The most common type of breathing apparatus in use with the fire and rescue

FAR LEFT Firefighters' gloves should provide excellent protection against heat and sharp objects, as well as affording good dexterity. *(Bristol Uniforms)*

LEFT Firefighters' boots can be made of leather or rubber. Regular maintenance ensures that boots are extremely comfortable and hard wearing. *(Bristol Uniforms)*

ABOVE A Scott Pro Pak breathing apparatus set with Scott Sight built-in thermal image camera. *(Alfie Martin)*

services uses compressed air carried in lightweight cylinders. A self-contained breathing apparatus set will consist of a facemask with a speech diaphragm, a one-way exhalation valve and a fitting for the air intake.

The compressed air cylinder is commonly made from an extruded aluminium liner coated with a carbon fibre wrap, both to give it integrity under pressure as well as protection from mechanical damage. Air is pumped into the cylinder and stored at pressures up to 3,000 pounds per square inch.

The cylinder is fixed to a body-moulded backplate, to which body harnesses are attached to enable a comfortable fit to the wearer. A system of high-pressure hoses connect the cylinder to a pressure-reducing valve, which regulates the air pressure through an additional flexible hose connected to a demand valve, which securely attaches to the facemask. Breathing apparatus sets use a principle of positive pressure, meaning that there is always a higher pressure in the facemask than atmospheric pressure. This prevents the ingress of potential contaminants that could occur if the pressure in the facemask was lower.

Other hoses connect to a pressure gauge (manual or digital) that allow the wearer to measure both the contents of their cylinder and monitor air use during activity. Breathing apparatus sets also incorporate a low-pressure warning whistle as a mechanical safety device that warns the wearer if they reach a pre-determined pressure. This alerts the wearer of the need to exit the building.

An automatic distress signal unit (ADSU) is incorporated into a breathing apparatus set to act as an alarm in the event that a wearer gets into difficulty and requires urgent assistance. The ADSU will be triggered automatically if a wearer becomes and remains inactive for an extended period of time; full alarm is preceded by an alert to reduce the risk of false activation. There is also a manual activation button that can be operated in an emergency. In both cases, once the ADSU has been activated it can only be silenced through the device key, which will be at the breathing apparatus entry control point attached to the personal tally allocated to the breathing apparatus set.

The activation of an ADSU at an incident immediately prompts a breathing apparatus emergency procedure. Firefighters at the scene will immediately initiate a rescue procedure to locate and retrieve the firefighter in distress. Emergency teams are also deployed and the breathing apparatus emergency is reported to fire control, who allocate an immediate increase in incident resources.

The breathing apparatus tally mentioned above is attached to the ADSU and in turn the breathing apparatus set. The tally is part of the control system used when breathing apparatus is in use. This is explained later in further detail.

Personal line

The breathing apparatus set personal line is designed to allow wearers to either attach themselves to a team member individually, or to a pre-laid guideline. The personal line is 6 metres long with one end fixed to the breathing apparatus harness. The full 6m (20ft) length is sub-divided into a 1.25m (4ft) section, with the remaining 4.75m (16ft) separated by a specialist two-stage clip. The free end of the line has a large snap-on clip attached to fix to a team member or to clip onto a guideline, enabling the wearer to follow this line without becoming detached.

Personal lines used between team members enable firefighters to remain physically attached to one another while widening the search capability. When used with a guideline it adds an additional level of safety to prevent disorientation in low-visibility scenarios.

Torch

Each breathing apparatus set will also be provided with a torch. Generally, these will be right-angled and attached to the breathing apparatus set by a removable clip. Being attached leaves a firefighter with a free hand.

The torches used as part of the breathing apparatus ensemble are designed to be intrinsically safe. This means that the device is designed to afford protection to facilitate its safe operation in hazardous areas, e.g. explosive atmospheres, by limiting any electrical and thermal energy that could potentially cause ignition.

Additional breathing apparatus features

Technical advances in breathing apparatus continue to bring new features and capability to this vital piece of firefighting equipment. Modern compressed air cylinders weigh a fraction of their predecessors and the use of micro-electronics has led to opportunities to integrate devices such as thermal imaging cameras and heads-up displays within the facemask of a breathing apparatus set.

Many breathing apparatus sets are fitted with telemetry equipment that provides information to the operator at the breathing apparatus entry control point. This allows the entry control operator to monitor safety critical information for personnel committed to high-risk environments and to assist in identifying a wearer in the event of them becoming trapped, lost or injured.

ABOVE A combined telemetry, air pressure gauge and automatic distress signal unit. *(Alfie Martin)*

Breathing apparatus entry control procedures

The necessity for breathing apparatus comes from the need for firefighters to enter irrespirable atmospheres, whether that is through the smoke or the presence of noxious gases. When this is the case, the breathing apparatus wearer is wholly reliant on the correct functioning of the breathing apparatus set and the ability to return to a safe environment on the completion of a task or before they have exhausted the contents of their compressed air cylinder.

In circumstances where a firefighter becomes trapped, lost, or in very rare cases where a breathing apparatus set malfunctions, any rescue efforts are time-critical.

RIGHT A breathing apparatus tally is allocated to ever firefighter who is designated to wear breathing apparatus. Details entered on the tally ensures that the firefighter has sufficient air to complete the task. *(Devon and Somerset FRS, Taunton)*

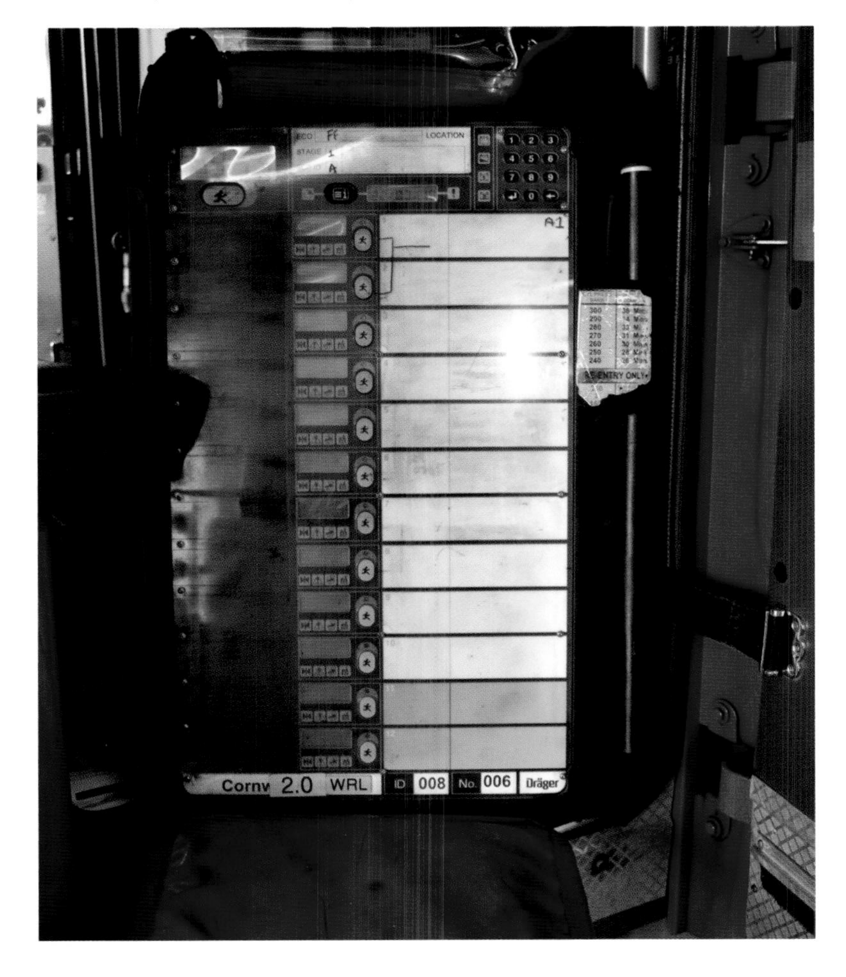

BELOW A breathing apparatus entry control board, with electronic telemetry, stowed in the crew cab of the fire appliance for easy access. *(Alfie Martin)*

The provision of strict accounting, control and emergency safety measures when firefighters are deployed in breathing apparatus is achieved through the use of breathing apparatus entry control procedures. These procedures form a national standard that is adopted and used by all UK fire and rescue services.

In any incident where breathing apparatus is deployed, a breathing apparatus entry control officer (BAECO) will be appointed, whose role will be to establish the entry control point (ECP) through which all firefighters will register before entry and return to on exit. The BAECO will operate a breathing apparatus entry control board which serves as a means of registering and accounting for all firefighters who have deployed through the ECP.

Each breathing apparatus set has its own unique identifying tally that is fixed to the key for the ADSU and that identifies the following information:

- The fire and rescue service
- The specific and unique number of the breathing apparatus set
- The name of the wearer
- The cylinder pressure
- The time they entered the risk area

The rear of the tally is used to record dosimetry readings in the event of an incident where the presence of radiation is known or suspected.

Information on the tally will be completed by the wearer as part of the standard test for the breathing apparatus set. The tally is then handed to the BAECO, who will place the tally in a breathing apparatus entry control board, where details of the time of whistle (required time of exit from the risk area) are calculated using the cylinder pressure, the location to which the wearer has been allocated as a part of a team and any other pertinent remarks.

The breathing apparatus entry control board is marked to provide its own identification details, provides space to record required information and will incorporate a clock and breathing apparatus set duration calculator.

Information is recorded on both the tally and the board through the use of a waterproof pen or pencil that can be erased once the individual wearers have been accounted for and left the risk area.

The BAECO is identified through a standard black and yellow chequered tabard.

In the event of greater numbers of personnel being deployed in breathing apparatus the management of breathing apparatus entry procedures is enhanced. The first step will be to move to stage two procedures, which sees the addition of supervisory staff and the provision of a dedicated emergency team. As resources further increase there is breathing apparatus main control procedure, which acts

THE IMPELLER

The impeller has been used as the basis for role markings in the fire and rescue service for many years. Depicted in silver, the impeller features in all fire service roles above that of crew manager.

RIGHT Watch Manager role markings. *(Cumbria FRS)*

and do not rely on internal valves or pistons. As the name suggests, this type of pump utilises centrifugal force (the force created when a revolving body causes items to fly out from the centre of the rotation). It is the same force that is used to 'spin-dry' clothes in a washing machine and that requires the rider on a roundabout to hold on tight.

An additional, smaller pump known as a primer is used when a pump is used to draw water from an open source. The primer acts to evacuate the air from within the pump, thereby creating a negative pressure or vacuum. This vacuum causes water to enter the pump system through atmospheric pressure. Once 'primed' and pumping water, the primer disengages as the passage of water operates like a siphon. The primer is only required in the event of the vacuum being lost.

The device in a centrifugal pump that creates the centrifugal force is called an impeller. It rotates on a shaft powered by the vehicle engine and has a series of semi-circular vanes spaced between the centre where water enters the impeller, reaching to its periphery. Water enters at the centre and is then channelled to the outside. The centrifugal force gives the water kinetic or velocity energy, throwing it outwards.

As water exits the impeller, it is ejected into the pump housing known as the volute. The volute slows the speed of the water considerably, converting the kinetic energy into pressure energy. The volute is cast in two parts and resembles the shape of a snail shell. It widens towards the outlet and channels the water under pressure towards the final pump outlet.

The speed of rotation of the impeller dictates the pressure at which the water leaves the pump, and this is controlled by the pump operator.

This type of pump is best suited to delivering water with high volumes but with relatively low pressure, for firefighting operations using larger diameter hose of 45mm or 70mm in size.

In order to supply the pressures required for an effective hose reel attack, the major pump is designed with two stages. This is a fairly simple engineering modification, which sees a second impeller fitted parallel to the first. The effect of this is that water from the first stage impeller enters the pump housing for the second stage, where another impeller further increases the water pressure. Water exits the second stage into the hose reel system at the much higher pressures needed for effective use.

ABOVE A light portable pump in use – the water is drawn into the pump through the grey hard suction hose and delivered to the firefighting branch through the red delivery hose. *(Cornwall FRS)*

RIGHT **An ejector pump – used in areas where fumes from traditional pumps would make it unsafe for firefighters to work.** *(Philip Martin)*

Light portable pumps

There are occasions where a means of pumping water remote from a fire appliance is necessary. This includes the extraction of water from remote water supplies, pumping out flooded properties and when water needs to be conveyed across open group inaccessible to a fire appliance.

A number of pumps are available in different types and sizes to achieve this within varying capacities and capabilities. A light portable pump will usually have a small petrol or diesel engine carried in a frame that can be easily transported to its required location. The pump itself may be similar to the major pump, although it will usually be of a simple single-stage variation.

The outlet of the pump will have a 70mm instantaneous valve and coupling and depending on the pump capacity, it may have a similar suction inlet. Smaller, more manoeuvrable pumps may use a smaller size suction hose.

Specialist pumps

While the major pump and the light portable pump are most commonly used within fire and rescue services, there are other pumps designed for specific operations where these would not be suitable or capable for the task required.

Ejector pump

There are occasions where the use of a pump using an internal combustion engine is not practical or feasible. For example, where the water that needs to be removed is too far away for the practical use of a suction hose, or where it is not safe to generate exhaust fumes from an engine. In these circumstances, there is an option to use an ejector pump.

Ejector pumps have no moving parts and use the flow of water from a pressurised source to propel fluids from source to exit point. Ejector pumps use the venturi principle, involving the creation of negative pressure to enable atmospheric pressure to fill an area where a vacuum has been created. Water entering this vacuum is then entrained into the propellant (water entering the pump under pressure) so that both propellant and fluid mix. This mixture then leaves the ejector pump through the outlet which may then be directed through a hose line.

BELOW **A fire hydrant located at the bottom of a hydrant pit. These are often found full of debris washed from roads and verges, making regular maintenance essential.** *(Alfie Martin)*

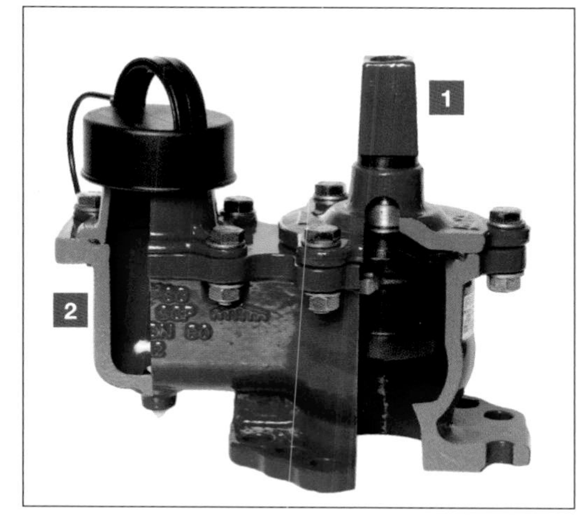

BELOW RIGHT **This cutaway of a fire hydrant shows the screw-down control valve (1) and the water discharge outlet (2). The black cap prevents mud and other debris entering the outlet.** *(Saint-Gobain PAM UK)*

Water supplies

Water supplies for firefighting come from two main types – pressure-fed mains and open water sources.

Fire hydrants

In urban areas and populated rural locations, mains water supplies provide a source of water to supplement firefighting operations. This water supply is the same one that provides the water to our homes and is drawn from the mains by fire hydrants. Fire hydrants are sub-surface valves fitted directly to the water mains that have a valve and outlet to which the fire service can connect hoses through a standpipe. The standpipe is controlled by a valve that is operated using a specific pattern key at the end of a shaft turned by means of a bar used to create a T key.

Fire hydrants are fitted within a tone or brick-built chamber beneath roads and pavements. They are denoted by a post and plate that gives details of the size of the water main and the distance of the hydrant from the post. The hydrant is covered by a heavy steel plate marked FH. This is lifted for use using the formed end of the bar that forms the T key. The hydrant post, plate and cover are painted bright yellow to make them easily visible and to differentiate them from other underground services.

The size of the main and the pressure of the water in the main will dictate the volume and pressure available for firefighting. This can vary significantly across locations.

Open water supplies

Open water supplies can be natural water sources such as streams, rivers, lakes, ponds and the sea, or man-made structures such as swimming pools, ornate ponds and reservoirs.

There are locations where emergency water supplies are provided through large static tanks, either above or below ground, and these may be found in remote rural areas as well as in industrial complexes.

Pumping from an open water source requires access to the supply using a pump to provide water under pressure, directly or indirectly to the scene of an incident.

LEFT A fire hydrant marker post provides firefighters with vital information about the water supply. The top number is the size of the main (in millimetres) and the bottom number indicates the distance to the hydrant (in metres). *(Shutterstock)*

BELOW The hose reel and branch allow firefighters to get water onto a fire extremely quickly, as everything is pre-connected and ready to use. *(Alfie Martin)*

ABOVE **Rolled hose in a fire appliance locker.** *(Alfie Martin)*

Firefighting hose

In order to get water (or foam) onto a fire it needs to travel from the water supply to the pump and then from there onto the fire. Different types of hose are required for different tasks and roles.

BELOW **Firefighters making up Dutch rolled hose.** *(Alfie Martin)*

High-pressure hose reels

In the UK, firefighters use hose reels extensively. A hose reel uses solid hose, typically 19mm (0.75in) in diameter, in 20m (66ft) lengths. These are usually installed in three 20m lengths on fixed drums on each side of a fire appliance, although this may be a single reel at the rear centre of the vehicle. At the end of the hose reel is a high-pressure branch that has a simple trigger operation and multiple jet and spray patterns.

High-pressure hose reels have been the preferred method for first attack for over 50 years. Being fixed and enabled for immediate use means they are quick to get to work. The flexibility afforded by a narrower-diameter hose makes the hose reel more manoeuvrable and the construction means hose reels are less likely to kink.

Lay flat hose

To get large volumes of water through a hose it needs to have a reasonable diameter. Most fire appliances carry 45mm and 70mm diameter hose in 25–30m (82–98ft) lengths. It would be impractical to carry the hose preformed to the diameter and it is therefore carry rolled as individual lengths.

Fire hose is made with a high-grade synthetic rubber lining to carry the water. This is surrounded by a bonded nylon woven outer sleeve, which protects the synthetic rubber while helping to prevent kinking when in use.

Each end of the hose line is fitted with a 70mm instantaneous coupling. The instantaneous couplings are the same across the UK Fire and Rescue Service, a result of the National Fire Service and post-war fire and rescue services.

The instantaneous couplings are attached to each end of the lay flat hose by means of woven cable. The male coupling is a simple single cast cylindrical inlet. The first length is inserted into the outlet of either a fire hydrant or a fire pump.

The female instantaneous coupling holds the male coupling in place through two spring-operated lugs that fit around a collar in the male coupling. A rubber washer is fitted in the female coupling and lengths of hose can be connected very simply and quickly. The connections operate very effectively under pressure.

The lugs on the female coupling also act as handles when the hose is being run out. A firefighter grips the lugs with each hand, rotating the hose and deploying as the firefighter moves

RIGHT Hose made up in the Cleveland Coil for use in high-rise firefighting. *(Alfie Martin)*

forward. When running out consecutive lengths of hose, a firefighter can kneel when they reach the end of a hose length then grasp the male coupling from the hose line of a colleague following on. The firefighter connects the two. This process can be replicated until the required length is achieved.

There are alternative ways in which delivery hose can be stowed ready for use. These include:

Dutch rolled

This is where the hose is rolled from a fold at the centre of the hose line. The hose is then stowed with the couplings resting one above the other. The hose can then be deployed swiftly by one person and may be advanced under pressure and passing water by a team.

Cleveland Coil

The Cleveland Coil is a method of stowing and carrying hose that was specifically designed to meet the needs of firefighting in high-rise buildings where space access is confined, particularly in access stairwells. The hose is stored by folding in a flaked style then secured for stowage and carrying using two retaining straps. This method of stowing the hose enables it to be carried over a firefighter's shoulder, leaving them with a free hand for manoeuvring.

Once the firefighter reaches the required location, connecting the Cleveland Coil enables the entire length of the hose to be deployed and charged with water, despite the restrictions of a confined space. This in turn allows firefighters to advance from an area of relative protection towards a fire with a charged and useable hose line.

Suction hose

Suction hose is used when a pump is drawing water from an open water supply such as a river or lake. Pumping from open water requires the creation of a vacuum. For this reason, the hose used between the pump and the water supply has to be rigid in order to withstand the negative pressure.

The procedure for using PPV – when performed correctly – proves very effective, with benefits to firefighting and firefighters through the rapid removal of heat, smoke and fire gases to create a sustainable working environment in which firefighters can navigate a building in tolerable conditions.

The need for specialist training and a competent crew is essential to ensure strict control and adherence to procedures. Conversely, a failure to operate within procedures could create a rapid worsening of a fire situation.

PPV involves a petrol-driven, electric or water-driven van, which is located outside of the affected building facing an opening. An exit route is formed elsewhere in the building, usually on the opposite side to the fan, and this outlet is protected by a covering firefighting jet. When operated, the fan will create a positive pressure within the building, forcing out the heat, smoke and fire gases. At the point of exit there is a need to apply water to both any flames forced out of the opening and any smoke and unburned fire gases, as the mixture of these with a plentiful supply of oxygen in the clean air could cause ignition leading to fire spread.

Rescue and extrication equipment

The tools used for rescue and extrication in the fire and rescue service have evolved significantly from the days of making do with equipment designed for use in vehicle repairs. Progress in developing hydraulic and electric power tools has led to the availability of a range of equipment specifically designed to accelerate the speed at which firefighters can access and release people who have become trapped in vehicles, machinery and other structures.

Hydraulic tools operate through the movement of hydraulic oil at extremely high pressures, causing components in tools to expand and contract. The strength of the material used, along with the extreme pressures, enable hydraulic tools to cut, spread and push metals and other materials exerting forces of over 100 tonnes.

ABOVE A range of hydraulic rescue tools – these give firefighters the ability to extricate trapped casualties quickly and safely. *(Shutterstock)*

RIGHT A dedicated hydraulic cutting tool – designed with sharp serrated blades to cut through metal. The hose lines from the hydraulic pump can be seen connected to the tool. *(Alfie Martin)*

BELOW A dedicated hydraulic spreading tool – designed with wide opening jaws to spread metal and create space. *(Devon and Somerset FRS, Minehead)*

The advance in battery technology has enabled rescue equipment manufacturers to use electrical power for devices ranging from reciprocating saws through to hydraulic tools. This has led to the availability of portable equipment for rescue use.

Hydraulic cutters

Hydraulic cutters used in rescue are capable of cutting through a wide range of metals and structures. For example, the cutters have the ability to cut through the pillars in a car in seconds, as well as being capable of shearing reinforced steel rods and metals used in building construction and machinery.

Hydraulic spreaders

Hydraulic spreaders can be used to create the space for initial access in rescue situations. With the ability to force objects and structures apart, spreaders can separate different elements of a vehicle, machinery or building in order to make space for extrication, or to gain access to use other tools.

The spreader force can also be used in the opposite direction to squeeze objects together for the purpose of space creation, or to enable the use of other equipment, such as cutters or powered saws.

The tips of many hydraulic spreaders can be

LEFT The Lukas battery powered combi-tool – designed to undertake both spreading and cutting of metal. *(LUKAS Hydraulik GmbH)*

BELOW An hydraulic rescue ram used to create space and provide firefighters and paramedics with better access to trapped casualties. *(Philip Martin)*

removed to enable the attachment of chains, creating the ability to use the pulling force of the spreaders.

Combination tools

Combination tools, better known as combi-tools, combine both the spreading and cutting ability of two separate devices into one multi-purpose hydraulic tool. The combi-tool uses the area of the spreader arms near the pivot point as the cutting element through serrated blades cast into each of the spreader arms.

Combi-tools provide an extremely useful first attack rescue capability.

Rescue rams

Rescue rams are extending tubes that can be used to lift, spread and stabilise vehicles, machinery and structures. The rams consist of a piston operated by hydraulics with reinforced stabilising plates at the base and end of the extension.

Used in conjunction with other hydraulic tools, rescue rams are very effective at creating the space to extricate casualties from damaged vehicles and machinery.

When used, rescue rams are usually supplemented with the use of solid chocks and wedges. These are put in place to stop any space closing up as the ram is removed, and

RIGHT Sharp end protection provides a safe working environment for firefighters, paramedics and the casualties by covering up sharp metal exposed during the extrication process. *(Devon and Somerset FRS, Taunton)*

can also be used to increase the ability of a ram to create a space wider than the standard range of operation by being used as a stage from which to operate.

Reciprocating saws

Battery-powered reciprocating saws offer a portable cutting capability that can be used on a variety of incidents. With the ability to use different blades, they can be used to cut a wide range of materials to assist in rescue situations.

Sharp end protection

The action of shearing and cutting metal by whatever means will leave sharp surfaces to structures. This presents a risk to both casualties and rescuers. Sharp end protection consists of

RIGHT High-pressure air bags in use to lift a vehicle. The air bag is protected from sharp metal or rough road conditions by wrapping it in a salvage sheet. *(Devon and Somerset FRS, Taunton)*

assorted sheets and sleeves made of Kevlar. This strong synthetic fibre provides a layer of protection over sharp objects, enabling rescue teams to work safely in rescue and extrication activities.

Airbags

The basic operation of an airbag uses the pressure created by expansion to lift, stabilise or spread an object or structure.

Airbags operate using compressed air from a cylinder, which is operated through a control valve assembly.

Low-pressure airbags are large cylindrical bags with a reinforced base and top. Once fully inflated, low-pressure airbags afford a reasonable level of stability, although as with other lifting tools, firefighters use chocks and blocks as objects are raised, in order to maintain stability throughout the operation.

High-pressure airbags are square and vary in size and lifting capacity. While the distance they can lift is less than that of low-pressure airbags, the lifting capacity is dramatically higher, with the largest being capable of lifting up to 67 tonnes.

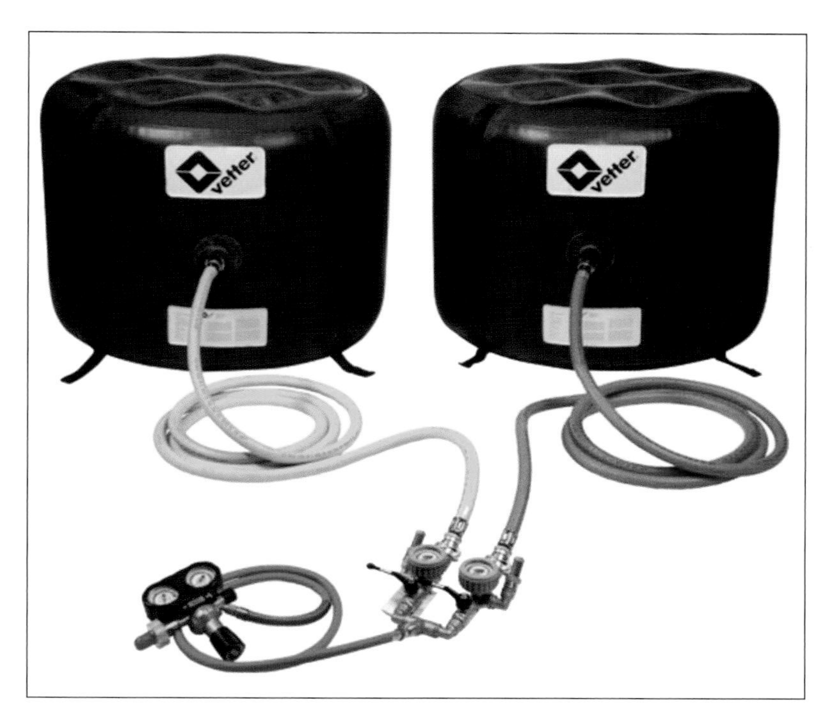

ABOVE Low-pressure airbags and controls. *(Vetter)*

Stabilisation

Stability is a key consideration in any rescue and extrication activity. Stabilisation involves securing the vehicle or structure for the safety of the casualty and firefighters. A failure to stabilise could be catastrophic for someone suffering serious spinal trauma, as well as leaving both casualties and rescuers vulnerable to injury through moving or falling objects.

LEFT Stab fast jacks in use at a road traffic collision to stabilise the vehicle. This allows firefighters and paramedics to work safely around the vehicle. *(Devon and Somerset FRS, Taunton)*

Stabilisation can be achieved through the use of chocks and blocks, jacks, expanding jacks and straps, and a wide range of other equipment.

The maintenance of stability requires constant monitoring during rescue operations as the movement or removal of metalwork, masonry or vehicle contents may be sufficient to render stabilising equipment useless unless it is adjusted to compensate for movement.

Rescue platforms

The operation of specialist rescue equipment is most safely achieved from a solid surface. In the case of cars and vans it is likely that the ground around the vehicle will provide this facility. The same cannot be said for large and heavy goods vehicles, coaches and mobile plant.

Rescue platforms are portable structures designed to provide the required height and working area to affect rescues in larger vehicles.

Glass management

Glass management forms part of a vehicle extrication process. The managed removal of glass at an early stage removes the risk of accidental breakages as the stresses on the vehicle change.

Side and rear windows will be retracted and removed through the use of a centre punch or other piercing tool, with any broken glass contained and removed from the scene of operations.

Some glass within a vehicle is not simply there for vision. In most cars the front screen will be integral to the strength of the vehicle. Increasingly, vehicles are supplied with a glass panoramic roof. These glass elements are manufactured and installed to be extremely strong, therefore removal or replacement requires specialist windscreen equipment.

In the case of a vehicle involved in a collision where the screen has been damaged, it may have maintained much of its structural integrity. Where the screen forms part of a vehicle entrapment, firefighters will need to either cut through or remove the screen to enable rescue.

Achieving this, in the absence of specialist windscreen equipment, requires a specialist glass saw. These are designed to create an initial hole in the screen before the saw blade can be inserted.

The material in glass screens can be harmful if inhaled or ingested. Firefighters will therefore provide eye protection and a dust mask to the casualty and wet the area to be cut to minimise the creation of airborne particles.

RIGHT Stab fast jacks can be deployed by a single firefighter ensuring the stabilisation process can be completed quickly. *(Devon and Somerset FRS, Taunton)*

LEFT A rescue platform in use during training, providing firefighters with a platform for safe working at height. *(Devon and Somerset FRS, Taunton)*

BELOW To create full access to casualties, firefighters will remove the roof of the vehicle. To achieve this, the windscreen requires cutting. *(Devon and Somerset FRS, Minehead)*

RIGHT **A steering wheel air bag protection system in use. This provides protection to the casualty, firefighters and paramedics where air bags have failed to deploy during the collision.** *(Devon and Somerset FRS, Minehead)*

Airbag cover

Vehicle airbags significantly reduce fatal and serious injuries in vehicle collisions. The isolation of the vehicle's battery by firefighters will significantly reduce the risk of an accidental deployment of an airbag, although there does remain a residual risk.

The risk of injury to firefighters through an accidental airbag deployment is greatest from the primary airbag fitted within the steering wheel.

Firefighters use an airbag protector to cover the steering wheel. This device is a circular Kevlar cover with nylon webbing straps to secure around and to the rear of the steering wheel. In the event that the airbag does deploy, it will be safely contained within the airbag protectors' structure.

Small gear

Small gear is the term used to describe the hand tools used by firefighters. General hand tools such as shovels, spades, drag forks, hammers, saws, pliers, spanners, crowbars, bolt croppers and screwdrivers will be included among the essential equipment carried on a fire appliance and these may be supplemented by specialist hand tools developed for fire and rescue situations.

Specialist items may include tools for breaking into secure premises and vehicles, and beaters for use in small grass fires and larger wildfires.

Emergency medical equipment

Fire appliances will generally carry some form of medical equipment. As a minimum, each appliance will carry a first aid kit.

Many fire appliances are equipped with medical trauma bags containing a selection of equipment for basic life support. This will include resuscitation equipment, airways to assist the breathing of an unconscious patient, medical oxygen and an automated external defibrillator (AED). Stretchers suitable for rescuing people from height may also be carried on a fire appliance.

OPPOSITE **Fire appliances carry a wide range of medical response equipment which can be used on members of the public as well as firefighters.** *(Devon and Somerset FRS)*

MAN
WA57
RT-61
25
DEVON & SOMERSET
FIRE & RESCUE SERVICE
FKV18474
FKV18474
FIRE
RESCUE
FIRST
AID

PREVENTING PROTECTING RESPONDING
www.cornwall.gov.uk/fire
UW 8480KG

Chapter Six

Fire appliances

Fire appliances, or fire engines as they are commonly known, are perhaps the most iconic tool in the fire and rescue service's extensive range of equipment. Fire appliances continue to turn heads, perhaps more than any other vehicle on the road, evoking childhood memories and prompting many a child to choose the role of a firefighter as their future career of choice.

OPPOSITE With the doors removed, the vast array of equipment and stowage configuration on a modern fire appliance can be clearly appreciated. *(Alfie Martin)*

Fire appliance history in the UK dates back hundreds of years, with evolution from simple hand-drawn pump and hose carts, through the era of heavy horse-drawn manual and steam pumps, through to motorised fire engines at the turn of the 19th century.

Modern fire appliances are designed with crew safety, performance, environmental considerations, optimal storage capability and flexibility in mind. A long way from the days of simply finding a means of transporting old technology to a powered chassis.

While a fire appliance capable of transporting a crew, a tank of water, a pump, ladders and a wide selection of both firefighting and rescue equipment remain the most common and versatile vehicles in any services fleet, there are a wide variety of specialist appliances in use that have been designed to meet the specific needs of modern risks across the UK.

In this section we will look at a range of fire appliances that represent the technology, capability and innovation used across the fire and rescue services in the UK.

There is wide-ranging terminology used by services to describe a type of fire appliance. In this section a general term may be used that differs from the terminology in a specific service. We will endeavour to explain what the general term used might also include.

TOP **A Northern Ireland FRS 4x4 Volvo appliance with a bumper mounted branch. This allows fires to be tackled while the vehicle is moving – ideal for wild land fires.** *(Emergency One)*

ABOVE **A Hampshire FRS Volvo appliance with multiple ladders and compressed air foam (CAFS) capability.** *(Steve Greenaway)*

RIGHT **A Royal Berkshire FRS appliance on Mercedes chassis.** *(Steve Greenaway)*

Major Rescue Pump

A Major Rescue Pump (MRP) is the Swiss army knife of fire appliances, carrying equipment to deal with most incidents that a fire and rescue service may reasonably expect to attend. Other titles may include Water Tender (abbreviated WrT), Water tender Ladder (WrL), Water tender Ladder Rescue (WrL/R), rescue pump and Pump Ladder (PL).

An MRP will typically be based on a 12.5 tonne commercial vehicle chassis. The crew cab will have the capability to carry a crew of up to six, and there will be a fixed water tank with a capacity of 1,800 litres (396 gallons). A major pump will be fitted to the rear of the appliance, driven through a power take-off from the road engine.

The rear of the appliance consists of storage lockers along each side of the appliance, and the roof will have the capability and facility to carry either a 10.5 or 13.5m (34 or 44ft) ladder, short extension ladder, roof ladder and potentially other large hand tools, all accessible from street level without the need to access the appliance roof.

The locker and crew cab will have tailored storage for the safe and accessible carriage for a range of firefighting, rescue and technical rescue equipment, along with first response medical equipment.

ABOVE Three compact fire appliances designed and built by Rosenbauer UK for Buckinghamshire FRS on MAN chassis. *(Rosenbauer UK)*

LEFT A Rosenbauer Compact Line demonstrator vehicle used to show fire and rescue services new innovation and technology in emergency response vehicles. *(Rosenbauer UK)*

ABOVE An IVECO Light Rescue Pump (LRP) in service with Devon and Somerset FRS. *(Philip Martin)*

The appliance will be fitted with visual and audible warning devices in the form of flashing lights and sirens, along with fixed and portable scene lighting.

Within the appliance cab, information and communications technology will include a multi-functional mobile data terminal with connectivity to data services/internet/intranet, mobile phone technology and digital radio transceiver with connectivity to fire control rooms and other emergency services.

The finish of the appliance will include high visibility markings on the front, back and both sides of the vehicle.

While the role of a major rescue pump will be similar across services in the UK, there is no single specification. This allows for flexibility to meet the specific needs identified by the services' own IRMP, although a number of services may work collaboratively to design and procure appliances to secure a common standard and achieve economies of scale.

Light Rescue Pumps

A Light Rescue Pump (LRP) is typically a smaller variation of an MRP. Commonly built on a 10-tonne commercial chassis, it will share many of the features of the MRP but with a reduction in capacities and capabilities, which will have been specified to meet the majority of needs of the area in its remit.

LRPs have evolved through the analysis of data and experience, which has shown that in more rural areas, a fully equipped MRP is not required. Reducing the equipment carried and using a smaller chassis holds a number of benefits for both the fire and rescue service and the firefighters who crew the appliance.

BELOW A trio of Mercedes Light Rescue Pumps (LRP) supplied to Cornwall FRS by Rosenbauer UK. *(Rosenbauer UK)*

Benefits will include:

- A reduction in the need for training on equipment that may receive little or no use with a consequent reduction in training time, testing and maintenance
- Smaller fire appliances are more manoeuvrable and provide more suitable access, particularly in more rural and congested areas where an MRP may have difficulty navigating narrow streets
- Smaller fire appliances will typically be less expensive to purchase, operate and maintain, along with better fuel economy and overall running costs

Even with reduced equipment, modern technology allows for smaller and highly capable equipment. For example, the use of battery-powered hydraulic rescue equipment will reduce stowage space and weight without any significant compromise in capability. In addition, the use of specialist firefighting media such as compressed air foam systems or cold cutting devices provide enhanced firefighting capability without the same need for a large water tank on the vehicle.

As with MRPs, a number of fire and rescue services have collaborated on the specification of LRPs, enabling services to purchase new appliances from a common specification.

Rapid Intervention Vehicles

The term 'Rapid Intervention Vehicles' (RIV) may also describe appliances termed 'targeted response vehicles', TRV or compact appliances.

Designed and built on a light van-sized commercial chassis, an RIV is a further step away from the MRP. Again, data analysis and operational experience has identified the scope for a smaller vehicle able to provide a first attack at an incident to provide an initial rapid response.

Similar to the LRP, these vehicles use modern technology to provide firefighters with equipment capable of dealing with a wide range of incidents. In many cases this capability will be sufficient to deal with the incident in its entirety, although operational procedures will have been designed to provide the subsequent additional support and capability of an LRP or MRP.

The identified and stated benefits of the RIV

LEFT A Devon and Somerset FRS IVECO Rapid Intervention Vehicle. *(Devon and Somerset FRS)*

ABOVE A Cornwall FRS 4x4 Major Rescue Pump. The front-mounted winch can be used for stabilising vehicles during extrication operations at road traffic collisions. *(Cornwall FRS)*

include but expand those of the LRP to deliver a vehicle capable of responding to and dealing with or mitigating risk independently or with the support of additional resources.

The reduced demands on training for firefighters make this appliance more viable and practical in areas where the risk may be low and incidence of fire and rescue incidents are minimal.

As well as being a capable appliance for initial response, the RIV is also a practical vehicle for providing backup to larger incidents. It is frequently the case that a large incident will demand large numbers of firefighters, but not necessarily fire appliances. The historic 'one size fits all' approach has meant that a fully kitted fire appliance has been used simply as a means of transporting personnel. The RIV approach provides an efficient, multi-use fire and rescue vehicle that is more suited to the demands and needs of modern communities, firefighters and incidents.

Off-road fire capability

Each design of pumping appliance can also be adapted to provide all-wheel-drive multi-terrain capability. This may be a requirement to meet the needs of local risk, particularly in remote rural locations.

High Reach Extendable Turret

High Reach Extendable Turret (HRET) appliances are becoming more common across the UK Fire and Rescue Service. The modern HRET concept was originally designed in the UK by fire appliance manufacturer Rosenbauer UK, in conjunction with Lancashire fire and rescue service.

This type of appliance is very similar to an MRP with the addition of a rotating and telescopic boom assembly, onto which is fitted a remote-controlled firefighting nozzle and a pointed firefighting and perforated firefighting nozzle, capable of piercing the exterior fabric of many commercial buildings.

The benefits of this type of appliance include the ability to commence offensive firefighting operations from the exterior of a building, without the need to unnecessarily commit internal firefighting crews.

The turret can also be fitted with a thermal imaging camera, closed circuit television camera and flood lighting. The whole mechanism can be controlled by a wireless remote-control unit.

LEFT A Hampshire FRS 4x4 fire appliance with front-mounted winch and bumper-mounted branch. *(Steve Greenaway)*

LEFT **A Lancashire FRS MAN AT-Stinger High Reach Extendable Turret appliance from Rosenbauer. It has the ability to pierce through into a compartment and extinguish a fire.** *(Rosenbauer UK)*

Airport Rescue and Firefighting Appliances

ARFF appliances are highly specialised vehicles built to meet the needs of the role, the varied terrain of an airport and the risks associated with aircraft incidents.

ARFF appliances need to compensate for the potential in delayed access to water supplies, requiring them to be self-sufficient in the initial and critical stages of an aircraft incident.

In the UK, the Civil Aviation Authority (CAA) set the minimum standards for response to an incident on an airport, reflecting the need for a rapid attendance in the event of an aircraft fire or crash. CAA standards detail the requirements for water capacity, acceleration, maximum speeds and ability to cross a range of terrains and ground clearances.

The presence of large amounts of aviation fuel, hydraulic oil and other flammable liquids require the ability to immediately discharge firefighting foams in high volumes on approach to and in attendance at an aircraft fire.

In order to reach an aircraft anywhere on the airport, an ARFF appliance will have all-wheel-drive capability. The capability of the ARFF appliance will reflect the nature and size of the risk. While a smaller airport may be able

BELOW **A Rosenbauer Panther 8x8 Airport Crash Rescue appliance. All four front wheels turn simultaneously to provide greater manoeuvrability.** *(Rosenbauer UK)*

to operate with smaller vehicles carrying less water and foam, major airports will demand large-capacity appliances with the power and agility to drive both on and off road. For this reason, manufacturers build ARFF appliances ranging from four-wheel drive to eight-wheel drive capability.

In order to provide the capacity and means to attack a fire with large volumes of foam and water while on the move, ARFF appliances have a remote-controlled high-volume monitor on the roof. The position of the monitor enables the crew to apply firefighting media as they approach the incident, to lay a foam blanket on any running fuel on approach, onto parts already affected by fire and as a layer of protection to the remainder of the aircraft.

Specialist ARFF appliances may have a high reach extendable turret with remote-control monitor and piercing nozzle, with others fitted with a hydraulic platform.

Aerial ladder platform

Access to height is a common requirement of the fire and rescue service for a range of operations. Historically, fire and rescue services were faced with a choice between a turntable ladder or hydraulic platform. Both vehicles had their own advantages and disadvantages.

The design and introduction of the aerial ladder platform brought the two capabilities together in a single vehicle, making it the favoured design for fire and rescue services.

An aerial ladder platform (ALP) consists of an arrangement of pivoted and telescopic booms powered by hydraulic arms mounted on a mechanical turntable. At the top (head) of the booms is a hydraulically articulated cage, which serves as a working platform, rescue stage and mounting for a fixed firefighting monitor. An extending ladder is secured parallel to the booms and the whole mechanism is fixed on a frame to which stabilising jacks are fixed. Also attached to the booms is a fixed system of pipework to convey water from the base to the monitor in the cage.

This ladder and platform assembly is fitted onto a commercial vehicle chassis and cab, along with bodywork to form lockers.

An ALP may be deployed in situations where access to a high level is required that is either impractical or not achievable through conventional ladders. The cage on the ALP can be used as a stable but moveable platform to perform rescues at height and to provide a position to apply large volumes of water onto

BELOW A Rosenbauer Volvo Aerial Ladder Platform (ALP) with jack legs fully extended lifting the vehicle wheels clear of the ground. This is done to create a totally level working platform. *(Rosenbauer UK)*

LEFT A Rosenbauer Volvo Turntable Ladder (TL) with a crew cage fitted at the head of the ladder. *(Rosenbauer UK)*

a fire from height, as well as in use for scene assessment and illumination.

The ladder on the side of the ALP booms can be used as access to height and for multiple rescues when the ladder is positioned and stationary.

ALPs range in working height from around 28m to 56m (92 to 184ft) in length. Some are capable of operating below ground level, e.g. beneath a bridge or high road access. As well as height elevation and 360° rotation of the turntable, additional articulation and rotation can be achieved through cage and final boom movement, providing an extremely manoeuvrable rescue and working platform.

Turntable ladder

A turntable ladder (TL) is an extendable high-access ladder mounted on a turntable. Similar to the ALP, the ladder and turntable assembly are fixed to a commercial chassis and cab, along with the stabilising jacks.

Some TLs are capable of having a cage fixed to the head of the ladder to provide a stable working platform and position for a monitor.

Hydraulic platform

A hydraulic platform (HP) is similar in design and operation to an ALP. Unlike the ALP, the HP does not have the ladder attached to the booms.

An HP will have a cage permanently attached to the top of the booms. This cage can be used as a working platform, as the outlet for a water tower, for lighting and in some cases for lifting.

BELOW Crews utilise a Simon Hydraulic Platform (HP) during firefighting operations at a fire involving a thatched roof. *(Devon and Somerset FRS, Minehead)*

ABOVE **A 4x4 Special Rescue Unit (SRU) from Devon and Somerset FRS.** *(Steve Greenaway)*

BELOW **Cornwall FRS Mercedes Rescue Tender (RT) complete with front mounted winch.** *(Alfie Martin)*

BOTTOM **A MAN Rescue Tender (RT) from the Devon and Somerset FRS.** *(Philip Martin)*

Specialist appliances

While each type of rescue pump will carry a wide range of equipment, there are occasions where specialist items for specific tasks are needed. Appliances designed to meet these needs are known as specialist appliances. This title captures a wide range of vehicles.

Rescue tender

A rescue tender (RT) is a specialist support appliance that will have an array of equipment designed for a range of rescue scenarios, most commonly used for extrications at road traffic collisions. While the specification for an RT will vary between services, they tend to share similar types of equipment and capability to deal with transport and industrial accidents. The chassis will often have a permanent heavy-duty winch attached, giving the vehicle the ability to move heavy objects.

The RT will carry hydraulic rescue equipment, capable of cutting and spreading vehicles and other metal structures. The provision of large hydraulic rams provides the capability to create space for rescues from large vehicles, rail locomotives and other structures.

Heavy-lifting equipment including high-pressure airbags capable of lifting up to 50 tonnes are common, along with a built-in compressor powered by the vehicle engine providing power for air tools.

Working platforms enable firefighters to work safely at height when they need to access the cabs of heavy and large goods vehicles.

A wide range of extendable jacks, blocks and wedges are carried for use in supporting unstable vehicles and structures during lifting and spreading operations.

Petrol-powered chainsaws and disc-cutters are sometimes necessary for cutting heavy-duty timber and stone structures. In addition, there may be an array of hand tools for use in rescue and other operations.

Services operating RTs will have them strategically located to meet their risk needs.

Technical Rescue Unit

Technical rescue encompasses working at height (and below surface), working in confined spaces and for a number of fire and rescue services, for the rescue of animals.

Technical rescue will often involve traversing off-road terrain and the transportation of equipment to a remote scene of operation. For this reason, many services will specify an all-wheel-drive vehicle for the role.

Technical rescue units will carry both the personal protective equipment for operators and the specialist equipment to fulfil rescues in some of the most challenging locations and circumstances.

Equipment will include rescue lines, harnesses and mechanisms for securing and directing lines over deep shafts, cliffs or other structures. They will also include rescue harnesses for both human and animal rescue and an array of pulleys, anchors and rope rescue devices.

Vehicles will potentially carry a crew, although it is also common for services to provide a secondary support vehicle, such as an all-wheel-drive pick-up truck, which can convey both equipment and personnel to the scene of operations.

Water Rescue Unit

Water rescue demands another range of specialist equipment in order to meet the needs of crews attending challenging incidents.

Equipment required will include additional personal protective equipment, a motorised boat, inflatable paths for use on mud and water, as well as rescue lines and other rescue equipment.

As well as operating on inland water courses and lakes, some of the most demanding water rescue incidents occur at times of inland flooding. In order to navigate affected areas, an all-terrain capability is essential for vehicles chosen for this role.

In the same way as a technical rescue unit, it is common to have a secondary all-terrain support vehicle.

ABOVE A 4x4 Special Rescue Unit (SRU), boat and trailer. The boat has just been recovered from the water following a training exercise. *(Andy Green)*

Command Unit

A command unit (CU) provides the focal point for incident and communication at a large incident. Command units are often manufactured through the conversion of a large panel van, although some services have larger coach-built units mounted on a lorry chassis.

While the size of the vehicle will determine the amount of equipment carried, most will share a similar basic design, incorporating information technology and providing

LEFT Surrey FRS Renault Command Unit (CU). The side and rear awnings provide additional space for command briefings and multi agency meetings. *(Primetech)*

ABOVE Northampton FRS and Northampton Constabulary shared Command Unit (CU). The side slide-out provides additional space within the vehicle and allows for a meeting room. *(The Northamptonshire FRS and Northamptonshire Constabulary Joint Operations Team and The Excelerate Group)*

communications with commanders on scene, access to information from their services command and mobilising system within the fire control room, access to the digital national Emergency Services Network (ESN) which provides communication with the resources of their service, neighbouring services and other emergency services. There is also the capability to access the service intranet, internet, capability to access remote CCTV, satellite communications, and links to download footage from police helicopters.

Larger command units may have been designed and purchased as a shared resource between two or more services (e.g. fire, police and ambulance). Where this is the case, the capability will be extended for each of the services' needs.

Command units are designed to act as the single point of contact at the scene of the incident. All communications to and from the incident will be channelled through the CU with the crew acting in a command support function. They will administer decision logs, analytical risk assessments and messages from and to the incident commander. They will also act to liaise between individual sector commanders and specialist officers at an incident. The CU will usually be the location for multi-agency briefings and where there is an attendance by more than one emergency service, command units will be co-located to provide shared support and communication.

The layout of the vehicle will be such that the incident commander can access incident information that can be displayed to assist in briefing and tasking of activity. Command units will also have some facility for on-site meetings.

Water carrier

The provision of a large water supply through a hydrant system or accessible open water supply is not always available, particularly in rural areas. A water carrier (WrC) is a large-capacity tanker designed for the purpose of carrying water for firefighting.

The vehicle itself can be similar to any commercial liquid-carrying tanker, although it will have fire and rescue service standard fittings to enable it to be filled and to empty its supply into fire and rescue service hose and equipment.

Water carriers will carry a collapsible dam capable of holding the capacity of the water tank. This enables crews to set up a static water supply near to the scene of operations and then travel to a suitable water supply to refill. This enables the water carrier crew to operate a shuttle between a water supply and the scene of operations.

The water carrier will also have the means for pumping water from an open water supply into the tank, hose and equipment to draw from hydrants and hose to supply the dam or another fire appliance.

LEFT Hampshire FRS Volkswagen Hazardous Materials and Environmental Protection Unit (HMEPU).
(Steve Greenaway)

Environmental Protection Unit

The role of the Environmental Protection Unit (EPU) is to respond to incidents involving a potential impact on the environment through contamination. Many services operate their EPUs in a modified panel van.

An EPU will carry equipment for the identification of hazardous materials, as well as the means to halt and contain spillages. Many fire and rescue services operate appliances with the support of the UK government's Environment Agency.

Equipment carried will include containers large enough to hold leaking drums or cannisters, specially absorbent and neutralising materials to contain spills and absorb contaminants, floating booms to prevent contaminants travelling through watercourses, inflatable bungs to block drains and other openings and special sealants to seal leaks.

Analytical equipment is also carried to enable the identification of a range of solid materials, liquids and gases, as well as specialist personal protection equipment for firefighters specially trained to deal with a hazardous materials (hazmat) incident.

Welfare Support Unit

Where an emergency incident involves crews staying on scene for a protracted period of time, there is a need to provide basic welfare facilities to enable crews to continue operating. Emergency scenes can be remote, with limited access to essential facilities, therefore services will have some means of providing this.

Welfare units may be self-contained vehicles or transferable modules that use multi-use prime movers to transport them.

Specialist dog transport

Dogs are used for a range of roles in the fire service, ranging from those trained to identify traces of accelerants (flammable liquids) as part of a fire investigation through to dogs trained to locate human scent as part of an urban search and rescue.

It is vital that these specially trained animals are transported safely and securely. The nature

BELOW The fire investigation dogs from Cornwall FRS have a specially adapted van to transport them, and their handler, to incidents. The dogs are affectionally classed as K9 assets.
(Cornwall FRS)

ABOVE A fireboat from London Fire Brigade undertaking a training exercise on the River Thames. The boat is equipped with a powerful pump capable of supplying a number of fixed monitors located on the deck. *(Shutterstock)*

of their role means that this may involve some lengthy road journeys.

Transport will normally be within a specially modified air-conditioned light van fitted with a secure cage and stowage for equipment and the welfare needs of the dog and handler.

Fire boats

As well as providing boats specifically for water rescue, where a fire and rescue service has a maritime or estuarial water risk, they may have a waterborne fire and rescue capability. London Fire Brigade operates a pontoon-based fire station on the River Thames, with crews staffing a number of craft capable of both providing a firefighting and rescue capability in the Thames as well as transportation of equipment along the waterway.

Other services operate vessels ranging from fully covered examples to rigid inflatables built to meet the needs of the specific risk areas. Fire boat crews are specially trained in maritime navigation and in dealing with maritime incidents.

BELOW This Fire and Rescue Service rigid inflatable boat is fitted with powerful outboard motors and is capable of carrying a crew of five firefighters. *(Cornwall FRS)*

National Resilience appliances

In order to provide a capability to respond to exceptional incidents resulting from natural and man-made disasters, as well as being prepared to deal with the threat of terrorist activity arising from chemical, biological, radiological, nuclear and explosive (CBRNE) incidents, the UK government funds a National Resilience Programme, which sees national assets in the form of vehicles, equipment and specialist capability, operated on behalf of the government by fire and rescue services under the management of the National Fire Chiefs Council (NFCC).

Appliances operated under this programme provide the UK with a response capability across the whole of England, Scotland, Wales

LEFT FIAT British Red Cross Fire and Emergency Support (FES) Unit. This unit provides firefighters with refreshments at large-scale or protracted incidents. *(Steve Greenaway)*

police, ambulance and other emergency service responders.

Salvation Army volunteers and officers provide this service to those working on the front line and within the community in emergencies ranging from large fires, widespread flooding and other weather-related events, missing persons searches and other occurrences.

The service is well-supported by the fire and rescue service and is often operated as a partnership between the two organisations at a local level.

LEFT Volunteers from the Salvation Army provide firefighters and first responders with refreshments at an incident. *(Salvation Army)*

MOTOROLA
USE LOGIN CSPG
Goto
Map
Menu
View
Enter STREET Name
Search
Results
GIS
Info
1 2 3 4 5 6 7 8 9
Q W E R T Y U I O P
A S D F G H J K L
Z X C V B N M , .
Enter
Func.
Menu
Space
Del.
Clear
CSPG
GPS (2)
Map
AIUS 24/2/19
STATUS 5 : MOBILE AVAILABLE
STATUS 14 : MOBILE TO STAND BY LOCATION

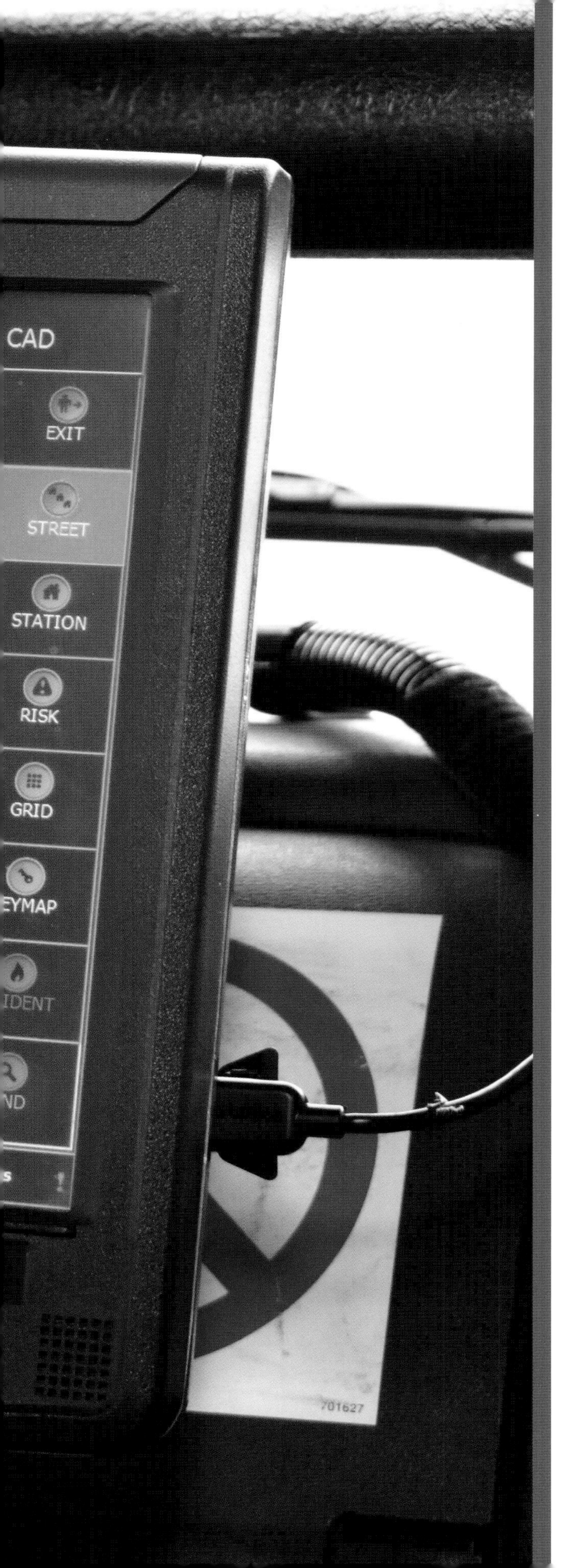

Chapter Seven

The anatomy of an emergency call

Emergency calls to the fire and rescue service can come via an automatic receiving centre or from another emergency service seeking assistance. All calls are directed through the relevant fire control room, but the most common way in which the emergency calls are made and received is through the emergency 999 system.

OPPOSITE A Mobile Data Terminal (MDT) fitted in the front of a fire appliance for retrieving risk-related data relevant to an incident. *(Freddie Martin)*

RIGHT Using a mobile telephone to dial 999 for emergency assistance. *(Alfie Martin)*

Calling 999

When you dial 999 (or 112), you speak to an emergency call handler from within a British Telecom (BT) dedicated emergency call centre. This operator will ask which service you require: either fire, ambulance, police or coastguard.

Using an automatic location system, the BT operator can identify your approximate location. In the case of a 999 call by landline, the operating system is linked to the database of addresses linked to phone numbers for a match. When calling from a mobile phone, the location is identified through the automatic recognition of the phone mast through which your mobile phone has connected to your supplier's network.

Having identified your location, the operator will seamlessly pass you through to the control room for the fire and rescue service closest to you.

Fire and Rescue Service control room

The operator in the fire and rescue control room will answer your call after transfer from the BT operator. The fire control operator will ask for information from you to determine the address and location. They will ask what the nature of the incident is and whether anyone is known or suspected to be trapped or in danger. They may also ask whether the incident is contained or developing. If the location is not clear they will ask for any landmarks or other

BELOW A Fire Control Operator's dispatch desk contains monitors and keyboards to access call data and risk information, telephones and a radio to communicate with officers and appliances. *(Cornwall FRS)*

Chapter Eight

Emergency incidents

This section will explore the range of incidents that fire and rescue services are called to, in addition to some notable examples that have occurred in the UK over recent years.

OPPOSITE A fire involving a large warehouse. Firefighter safety is vital at this type of incident. *(Shutterstock)*

Fires

While the chemistry behind fire remains the same, modern materials, structures and vehicles that are in use and being developed are very different. Changes that bring about greater efficiencies in modern building designs can impact on both the way fires are extinguished and on the conditions in which firefighters have to operate. This change has had to be met through changes to firefighting techniques and equipment in order to maintain the safety and capabilities of the modern firefighter. The basic principles of firefighting remain the same: to save lives, reduce and mitigate damage and to protect the environment.

The demand for housing and freedoms of design through modern materials continues to see commercial buildings and the homes we live in changing in the way they are designed, constructed and built. No longer are shapes and sizes constrained by angular features, and through smart engineering, structures can be built in places that were inconceivable in past decades.

The types of workplaces, homes and locations people live in is hugely varied. There have been significant advances in the safety of homes brought about through changes to building regulations and other legislation that has seen greater design flexibility to allow engineered solutions to compensate for a reduction in traditional safety features. These include improved compartmentation, requirement for smoke alarms, materials that through relevant standards should prevent the spread of fire and building designs that require the provision of safe means of escape.

BELOW **Firefighters tackle a well-developed house fire using a hand controlled branch and triple extension ladder. A safety cordon has been put in place to keep firefighters clear of falling debris.** *(Devon and Somerset FRS)*

In homes, the availability, reliability and reduced cost of smoke alarms, along with extensive programmes of fitting and community safety education by fire and rescue services, have all contributed to a reduction of fire deaths in the UK.

While people are generally safer in the built environment and in their homes, the challenges to firefighters has increased through some the modern building practices. Thermal insulation, essential for energy efficient homes, will affect the way a fire behaves. The risk of backdraughts and flashovers increases where natural ventilation does not occur. Building security makes access to buildings for firefighting more difficult and a practice as seemingly simple as placing cables in plastic trunking has led to firefighter fatalities. Furthermore, as car ownership has risen, simply reaching the scene of an incident has become more difficult with access for fire appliances regularly blocked or restricted by poor parking.

Fires can occur in and involve almost anything. We will look at the most common types of fire attended by the fire and rescue service.

Dwellings

The term 'dwelling' is used to describe a place where people live and sleep. Modern homes come in a wide range of designs, shapes and sizes, and sadly, while numbers have fallen, more people still die as a result of a fire in their home than any other type of building.

Dealing with a fire in someone's home will differ on almost every occasion. Security, lifestyle, health and mobility of occupiers will be just some of the factors that will have an impact on the nature of risks and accessibility for responding firefighters.

Good housekeeping, regular maintenance and servicing of installations and apparatus will ensure risks are reduced, but where any or all of those factors are neglected then risks will increase.

The majority of fires in the home continue to originate from the kitchen, with unattended cooking continuing to pose a high-level risk. A reduction in the use of cooking oil in pans and the increase in the use of purpose-made deep fat fryers has led to a decrease in fire deaths, injuries and the severe damage associated with chip pan fires. Electrical 'white goods' such as fridges, freezers, washing machines and tumble dryers continue to be a regular cause of incidents.

Technology such as auto-shut-off devices, smoke and heat detectors, and water-misting fire suppression systems can significantly reduce the risks and consequences of fires in the kitchen. Manufacturers have a significant role to play through ensuring that identified product faults lead to modifications, information to purchasers and product recall.

LEFT Firefighters using an Aerial Ladder Platform (ALP) at a fire involving a high-rise building. The versatility of these vehicles allow them to be used for undertaking rescues and firefighting. *(Cornwall FRS)*

High-rise dwellings

High-rise buildings are designed to resist fire, stop the spread of smoke and provide a safe means of escape for occupants. In principle, an individual dwelling in a high-rise building should not present any greater risk to residents than that of a low-rise building. Individual dwellings should be capable of containing a fire, while those adjacent should have structural protection to prevent fire spread.

Construction features common to high-rise buildings such as shared access, distances of travel and height, building exposure to the elements and potential damage to the integrity of construction between common areas, such as floors, walls and building services, do increase difficulties for firefighters and occupants when a fire does occur.

Modernisation projects to improve the aesthetic appearance of older high-rise properties and/or to improve thermal efficiency has seen the application of external cladding. This has been a contributory factor to rapid fire spread on the outside of buildings throughout the UK and wider world. From a firefighting perspective, this can create significant risks to firefighter safety and challenges the capability of firefighting equipment.

Additional fire protection features are provided within high-rise buildings, including fire detection and alarm systems, emergency lighting on escape routes and fixed water supply systems that enable firefighters to supply water to levels where a fire has occurred.

In some buildings there may be engineered solutions to deal with controlling and ventilating smoke, fire suppression systems such as sprinklers and water-misting systems. Fire suppression systems are of particular note as a safety feature, as they not only detect a fire but also operate to contain and extinguish it.

SPRINKLER SYSTEMS

When a sprinkler system operates, each sprinkler head will go off individually. It is not like the movies – they do not all go off – meaning water damage is minimal. Also, while the technology is over 100 years old, fire sprinklers very rarely go off accidentally.

Basements

A fire in a basement will present firefighters with unique challenges. Basements, sometimes known as cellars, can be shallow with just one or two levels below ground. Deep and often complex basements may have ten or more below-ground levels.

In a shallow basement, it may only be the floor separating the basement from the ground storey that is constructed as a fire-resisting compartment floor. Below this level, floors may not be fire separated from the basement floor above.

Fires in a basement may burn unseen for some time and may be difficult to locate upon the arrival of firefighting crews. Limited and difficult access can make a fire in a basement a challenge. Routes may be either inside or outside the building and may be limited to one single access point. There is also a likelihood that the route into a basement will also be the route for smoke to exit the building.

Ventilation and natural light may be very limited within a basement, although some may be provided through pavement lights and doors. In a basement, the risk of rapid fire development or backdraught is greater if ventilation is inappropriately commenced during firefighting operations.

As a contained space with highly insulated walls, a fire in a basement can develop rapidly, creating extremely high temperatures.

BELOW An office building severely damaged by fire. Fires in commercial buildings can have a devastating effect on a business. *(Shutterstock)*

Commercial buildings and plants

The commercial building and plant category can include anything from a small shop to a multi-acre chemical plant. Each will present multiple and sometimes complex risks, demanding that firefighters possess specialist skills and knowledge to deal with them.

A well-managed building (of any type) will have a good fire safety management plan which ensures fire exits remain clear, fire alarm systems are regularly tested and staff hold a basic knowledge of fire training. The absence of good management will often lead to poor housekeeping and ill-prepared staff, leading to an increase in risk both of a fire starting and inappropriate action should a fire occur.

While some smaller commercial buildings will present some generic risks, in the case of more complex buildings or those where there is high-risk storage or processes, it is essential that risk details are recorded and maintained for the information of responding crews.

This information will be kept on a database accessible to firefighters while en route to and at the incident. In order to gather and maintain this information, firefighters are entitled, with reasonable notice, to access any premises through powers conferred under the Fire and Rescue Services Act 2004.

Offices

Offices can range from a single room in a converted building to purpose-built facilities that accommodate potentially thousands of staff. Many office buildings will be shared between companies, which makes an accurate assessment of life risk, numbers of people in a building and risks created by items stored within the building difficult.

The nature of operations in a building will have an impact on the fire and rescue service response and access. Where the operations within the building are highly sensitive or involve large financial transactions, the offices are likely to be highly secure both on initial entry and throughout the internal parts of the building.

Shops and shopping centres

Shops and shopping centres come in a significant range of shapes and sizes and may be located almost anywhere, from a local

convenience store to purpose-built large-scale shopping malls housing hundreds of shops in a covered complex.

The relatively static numbers of staff in a shop or shopping centre will be swelled during busy periods by customers representing all of society. In addition, there will also be additional maintenance or service staff on the site.

Access to and around shops can be challenging for firefighters, particularly in busy periods and with the fluid nature of occupancy and type of business, a fire in a shop can present responding crews with difficulty in identifying the types of risk, layout and number of occupants.

Warehouses

Warehouses are key to the timely storage and distribution of millions of tonnes of goods on a daily basis. The type and scale of a business will determine the size and volume of warehouses, which can range from a single product type, i.e. plumbing supplies, to buildings in excess of 80,000m^2 (which equates to the same as eight football pitches). Some of the largest will be regional retail warehouses stocking bulk quantities of the entire range of items in a supermarket or department store.

The flow of deliveries to and from warehouses will often be 24 hours daily and numbers of staff may fluctuate during that time.

Modern warehouses will have high-level storage racking, which is managed and operated by fork-lift trucks. These may be fully automated and therefore driverless, operating through wireless telemetry which enables goods to be picked for specific loads.

Construction and demolition sites

The period of construction of a building will be the time between planning approval and building completion, whether it is a new-build or during works to expand or improve the building design and layout. With space for development at a premium or, where a business needs to adapt its accommodation, construction will follow an initial demolition to clear the site.

Construction and demolition sites present firefighters with a range of risks and challenges. In both cases, there is legislation that must be adhered to for the safe management of projects. The UK Health and Safety at Work Act is the primary legislation designed to ensure that risks at construction and demolition sites are kept to a minimum. In addition, fire safety legislation also applies.

During large-scale construction, developers are required to maintain risk information in accordance with construction (design and management) regulations. This information must be made available to responding crews.

ABOVE Firefighters at the scene of a major warehouse fire. Identification tabards are used to indicate firefighters undertaking specialist functions at the incident.
(Steve Greenaway)

RIGHT Fires in buildings under construction can difficult to contain due the lack of fire-protection measures during the building phase. *(Shutterstock)*

A properly managed and secured site will not generally create a situation requiring a fire and rescue service response, but, of course, there is a risk of poor management, unprofessional workmanship, poor-quality materials, a deliberate act of vandalism or fire-setting.

In the case of large-scale construction or demolition the fire and rescue service will have been engaged at the planning stage, which allows for pre-planning. As sites develop, there will be changes in the type and location of risks and access arrangements. In order to maintain knowledge of these areas, the fire and rescue service may arrange site visits throughout the work programme. In smaller-scale sites and when alterations are being carried out, these may be unknown to fire and rescue services.

During the demolition phase for a building it is necessary for firefighters to consider unseen and unrecorded risks. The presence of hazardous substances can come from older building materials such as asbestos, and where materials used in processes during the building's use have been left on site or incorrectly disposed of within buildings, in outside storage areas or underground. Any of these unknown materials can present a risk to firefighters.

BELOW Tackling fires in waste-processing facilities is often hampered by the presence of hazardous materials and specialist waste. *(Cornwall FRS)*

Engineering workshops

Engineering workshops exist in various sizes and locations. Many major manufacturers get their individual components from a whole host of smaller manufacturers across the UK.

As an example, an aircraft manufacturer will carry out final construction at a large site where it is likely that the airframe and engines will be fitted and from where the final product will emerge. The sub-components may well have come from a smaller facility, which could involve a single person operation through to a medium-sized factory employing 50 or more people. These can operate in a small space next to the operator's home or a medium-sized commercial building on an industrial site.

The main construction site for large industry may be vast. As an example, a car manufacturing site may cover many square miles with multiple buildings housing the manufacturing facility for sub-components, warehousing, offices and test facilities. There is likely to be a high level of automation in each of the engineering facilities, along with operators and final finishers.

At the other end of the scale in size, but far more common, will be car and commercial garages. These represent some similar risks, albeit on a smaller scale, and will be present in most localities. Engineering processes, manufacturing machinery, hazardous materials and compressed gases used in engineering can also present a risk to firefighters.

Factories

A factory may be similar to a large engineering or industrial plant. Factories represent an industrial site in their own right with warehousing, manufacturing, offices and distribution buildings within a large complex.

A factory may take raw materials and produce formed products such as steel or plastic. Mass-produced consumables, chemicals and machinery can be produced over prolonged product lifetimes. Modern factories are generally highly automated, replicating actions and processes thousands of times a day.

A fire in a factory can be complex to contain and extinguish. The complex nature of the building and machinery along with hazards presented by chemicals, compressed gases and unknown materials make pre-planning essential for responding crews.

Food processing and packaging

Food processing and packaging is a major industry, with sites across the whole of the UK. The need for impeccable standards of hygiene within highly insulated facilities has seen styles of construction evolve rapidly over recent years. Lightweight construction techniques, composite large insulated panels and an ever-increasing customer demand have all contributed to the development of large facilities that service an estimated 30 or more food manufacturing and processing industries.

Food processing ranges from hot cooking in the production of bread, baking and ready meals through to refrigeration and chilled areas for the production of vegetable, fish and meat processing.

Landfill sites

Landfill sites are large areas of managed outdoor areas that are used to deposit a wide range of waste products in a way that is safe to the environment. The nature of the waste will be dependent on what the site is licensed to accept. Sites can be many acres in size. When the site, or parts thereof, become full, the area is landscaped and the site becomes almost unseen.

In order to deal with the fluids and gases that may be created and generated by biodegradable products, landfill sites will have a network of pipes and drains to safely transport gases and liquids released by the waste. The gases can include methane and carbon dioxide. In the most efficient sites this gas is used and converted to useable energy. Contaminated liquids are collected and treated to prevent entry into watercourses.

The production of flammable gases from landfill sites can present problems for the fire and rescue service, where sites are poorly managed

ABOVE An aerial view of a major fire in a food-preparation facility. The construction of these buildings makes firefighting difficult and dangerous. *(NPAS)*

BELOW Fires in landfill sites can burn for very long periods. Firefighters often spend days tackling these fires. *(Western Gazette, Yeovil)*

ABOVE Firefighters in attendance at a fire in a recycling plant. *(Cornwall FRS)*

BELOW Firefighters using a triple extension ladder to access a railway carriage. Breathing apparatus and a hose reel jet are being deployed. *(Devon and Somerset FRS, Minehead)*

or effective means of extraction are not in place. In addition, if a fire does ignite in a landfill site it can be difficult to locate and extinguish with risks to responding crews created by the unknown nature of items that are involved.

Recycling sites

As more and more recyclable products are produced, the need for recycling sites has expanded. Many waste products are collected and recycled, including food products such as waste cooking oil, food containers and a range of other products such as metal, tyres and plastics.

While many recycling sites are licenced and well managed, there are others that are not. There have been many examples of fires involving unlicensed recycling sites where thousands of tonnes of products have been accumulated with no onward facility for safe disposal. Incidents involving fires in unauthorised recycling sites can create environmental issues on a very large scale and require large numbers of fire and rescue resources, often over extended periods.

Large transport hubs

Local, national and international travel will commonly require transit through a large transport hub such as a railway station, airport or dock. Each of these facilities will present unique issues for fire and rescue services.

Railway stations and infrastructure

The United Kingdom rail network is critical for both public transport and the transportation of goods. Rail travel has proven to be safe over many years, with risks well managed. When an incident does occur on the network, well-practised procedures ensure a swift response involving the rail network providers, rail operators and the emergency services.

Railway stations range from small unstaffed and often remote facilities to large hubs that serve significant numbers of people. These individuals may be starting or ending their journey, changing from one train to another or simply passing through.

Large railway stations will often operate not only as transport hubs but may also include a range of commercial premises including shops, restaurants, bars and offices. Numbers of people using railway stations will fluctuate, with many thousands of people transiting during peak periods.

In locations where there are large tunnels, rail operators and local fire and rescue services may have highly specialised equipment to provide both transport and equipment to the scene in the event of an incident.

Subsurface (underground) railways and stations present additional challenges to fire and rescue services. While emergency arrangements are well practised, with detailed inter-agency response plans, specific training for rail staff and inbuilt protection measures, an incident in a subsurface railway will present specific and challenging issues, which will include access, evacuation of the public and rail staff, underground communications, lighting, movement of equipment and ventilation.

The nature of risk for the fire and rescue service and other emergency responders requires close liaison with rail operators both in pre-planning and response to any incident.

Ship fires

Ships constitute the largest form of transport for goods to and from the UK, with an estimated 95% of all imports and exports being transported by sea. The country's manufacturing sector imports much of the raw material to make products that are then in turn exported by sea as finished products.

Shipping brings in much of the oil and gas used in the energy sector. Consumer products including food, medicine, cars and electrical products are also all part of the bulk transported by sea. It is estimated that annually around 65 million passengers use sea transport for business and pleasure travel.

With such a diverse cargo carried on increasingly large and complex vessels, the risks can be compared to large ground-based storage, work and leisure buildings with the complications of access, egress and construction that are unique to ships.

The area of responsibility for fire authorities to make firefighting provision in England, Wales and Northern Ireland generally extends to the mean low water mark at ordinary tide. That may include estuarial waters up to a prominent water mark. Beyond this area, there is no requirement or support through central government for the provision of a firefighting response.

Fires on board a ship alongside do fall within the responsibility of fire and rescue services. Where crews may be required to attend such an incident, they receive specialist marine firefighting training.

Ships are fitted with a range of fixed firefighting equipment and will be compartmented by steel doors. In the event of a fire on board a ship, the first response will be through the ship's crew who will have received basic firefighting training.

ABOVE Fires on ships are dangerous and challenge the fitness of firefighters. The use of multi-agency resources is required for firefighters to gain access. Here, a lifeboat from the RNLI has been used to transport firefighters to the incident. *(Cornwall FRS)*

A fire on a ship represents unique hazards through the materials and design of construction. Even when alongside, the access and egress to a ship will present difficulties. With the majority of a ship being metal, fire spread through conduction is common and internal ventilation systems can assist in unseen fire spread. The absence of external ventilation across the majority of vertical surfaces will lead to an intense build-up of heat, creating extremely punishing working conditions.

Firefighters responding to ship fires will be required to maintain knowledge and experience in the principles of ship construction. They will also need to be able to identify and use the ship's fixed firefighting, fire protection systems and understand the firefighting medium they use.

Incident command and control will be complex for a large or protracted incident and teams will need to operate from forward control points.

Water supplies for firefighting alongside need careful management; the monitoring of the ship's stability is critical when water is being used for firefighting.

For the incident commander, liaison between with the ship's captain, crew, the harbour master and other authorities will be essential throughout firefighting operations.

Airports

Air transport continues to increase in popularity and be used nationally and internationally. The UK has a number of small airfields, regional

RIGHT Firefighters tackle a fire involving an aircraft away from the airfield. Foam is being used to deal with the flammable fuel fire. *(Cornwall FRS)*

BELOW A fire involving a combine harvester. Large amounts of smoke are produced due to the size of the tyres on the vehicle. *(Devon and Somerset FRS, Minehead)*

BELOW RIGHT Agricultural buildings have lightweight roofs which disintegrate very quickly during a fire. Fires in agricultural buildings are often monitored and left to burn out. *(Devon and Somerset FRS, Minehead)*

airports and international airports serving the worldwide population.

As a transport hub, airports support a huge range of operations beyond the take-off and landing operations. These will include; air traffic control, aircraft maintenance, on-ground movement, bulk fuel storage and refuelling, baggage handling, public transport, security, customs control and administration. Airport buildings act as retail centres with shops, restaurants and bars.

Air travel continues to be very safe, although the potential for an incident has to be treated seriously. Airports are required to provide a provision for fire and rescue response within minutes of an incident involving an aircraft on the airport. This forms part of the categorisation and operating ability for the airport.

Airport fire and rescue services are designed to operate an immediate response and fire attack for an aircraft in distress. Local authority fire and rescue services provide a response to back up and support aircraft firefighting crews and are therefore required to maintain the knowledge, competence and skills to deal with an aircraft incident.

Rural firefighting

Farm fires

Fires on farms and other agricultural areas can present some unique challenges. The nature of farming will often mean that farm buildings are in remote areas and that the presence of large numbers of livestock needs special management. Mass storage of dried hay and

SPONTANEOUS IGNITION AND COMBUSTION

Factors such as weather during harvesting, outside temperatures and ventilation can create the right circumstances for spontaneous ignition and combustion to occur. Spontaneous ignition occurs through heat generated by an exothermic reaction as a result of the activity of micro-organisms within hay and straw stacks. When the heat causes the temperature to reach the point of auto-ignition, a further phenomenon known as thermal runaway can occur, resulting in the fire developing to full combustion.

ABOVE Fire in heritage buildings are an increasing problem for firefighters due to the lack of fire protection measures. Here, a fire is consuming the bell tower of a church. *(Steve Greenaway)*

other animal feeds and straw used for bedding can also provide large amounts of easily combustible fuel.

Heritage buildings

Part of the UK's rich heritage is maintained through its many historic buildings and sites. The value of our heritage cannot be measured in financial terms alone. A fire loss can destroy irreplaceable buildings and artefacts.

Heritage buildings act as time capsules, holding the secrets and style of historic building design. They often maintain the contents of their era, ranging from priceless art to furniture and features dating back many hundreds of years.

The historic nature of construction in heritage buildings means that internal fire protection will be lacking. While this is not impossible to resolve without impacting on the original features, it can be a costly and complex project to achieve.

In historic buildings that are workplaces, such as buildings open to the paying public and those used for businesses, this is a legal requirement, therefore the appropriate protection measures will be in place. This is not the case for historic private dwellings and these can therefore become vulnerable to serious destruction in the event of a fire.

Many heritage buildings have well-planned and managed fire prevention regimes. Those held in trust of the nation will generally have a fire pre-plan which provides guidance to both staff and attending firefighters to assist in the salvage of the most important items contained in the building as well as any particular parts or elements of the building that need to be protected.

Operational difficulties associated with historic buildings will include non-fire-resisting walls, unseen voids between walls and floors that allow fire and smoke to spread unseen and wooden doors that offer limited protection for fire spread. Thatched roofs are effective at repelling the elements, making them difficult to penetrate in a fire, but are inherently combustible.

Many heritage sites and buildings will also be in remote areas, which may entail extended travel times for firefighters. They also may not have access to a good water supply. Fire and rescue service personnel will know of the local major heritage sites and will also have site-specific risk information to assist in the management of fire incidents.

Wildfires

The National Operational Guidance (NOG) for the UK Fire and Rescue Service defines a wildfire as 'any uncontrolled vegetation fire where a decision or action is needed about its suppression'.

The risk of wildfire varies across different areas and regions of the UK and the incidence of wildfires is generally linked to sustained periods of dry and warm weather.

Large areas of moorland present the greatest wildfire risk, with the type and size of vegetation, the provision of any maintenance regime and soil type all factoring as potential risks.

In the UK, wildfires tend to be most prolific in summer, where wildland has been without

ABOVE A firefighter tackling a developing wildfire using a fire beater and wearing a respirator for protection against the particles generated by the fire. *(Devon and Somerset FRS, Minehead)*

rain for a long period. If temperatures rise and remain high for extended periods, vegetation dries out, which creates warm dry fuel in large volumes. The introduction of an ignition source can lead to the rapid – and for some time unseen – development of an extensive fire front.

The rural nature of wildland will often require extended travel for responding fire crews. Faced with a developing fire and absence of easily accessible water supplies, this type of incident will rapidly draw on significant fire and rescue resources.

While an out-of-control wildfire is potentially devastating, planned and controlled burns are frequently part of wildland management plans. Pre-planning to ensure the protection of dwellings, heritage and areas of outstanding natural beauty is critical for fire and rescue services, as is the need to understand the best locations for creating firebreaks.

The potential environmental impact of a wildfire extends beyond the destruction of vegetation and wildlife and can have an impact on watercourses, as well as the long-term regeneration of an area. Where a wildfire extends to an area near population, there are also risks to the infrastructure if electricity supplies or communications lines are damaged. Societal impact can include the need to evacuate areas and can extend the need to consider the potential impact on air traffic.

In many areas, the fire and rescue service are key partners in the management of wildland, a relationship that can often see the sharing of resources, equipment, knowledge and operational information.

RIGHT Trench rescues are extremely dangerous and firefighters must ensure they follow safety procedures to achieve a safe and successful outcome. Regular training scenarios ensure firefighters maintain their skills in this specialist area. *(Shutterstock)*

Technical rescue

The origins of the modern fire and rescue service were focussed on fire. Through the passage of time that role has changed to encompass a wider rescue role that continues to develop.

The experience of UK fire services in the Second World War seems to have been the catalyst to this change with the bitter and extensive experiences of post-war bombings being the testing ground for firefighter evolution.

Technical rescue is the term commonly used to describe procedures requiring tools and skills beyond the standard range of fire and rescue service operational equipment.

Technical rescue scenarios will include vehicle and machinery rescue, rope rescue, trench rescue, mines rescue, confined space rescue, structural collapse rescue, water rescue, and urban search and rescue. The latter often sees UK firefighters operating internationally.

Vehicle rescue

Road traffic collisions have become an increasingly regular incident for the fire and rescue service. As vehicular traffic has increased and modern vehicles have evolved, the challenges of swiftly and safely getting people who become trapped in their vehicles out have increased. Thankfully, through the evolution of specialist road rescue equipment

routes of entry and exit from contaminated areas, ensuring members of the public and site workers are not exposed to undue risk.

Fire and rescue services carry equipment that enable crews to be decontaminated upon leaving a hazardous area, can identify unknown substances, and contain or dilute spillages and gaseous releases.

Where there has been potential contamination into public areas, firefighters and specialist hazardous materials officers are able to advise on measures to minimise risk to public health.

In the event of a major hazardous substance release, the fire and rescue service will facilitate the establishment of a specialist technical advisory group (STAC) that will draw together specialists from a range of disciplines including industry, environmental agencies and public health.

Emergency medical response

In the UK, the Emergency Medical Service forms part of the National Health Service (NHS) response. Fire and rescue services have always worked effectively alongside ambulance colleagues at a range of incidents, and firefighters have traditionally been trained to a basic level of first aid and casualty care.

In response to the increasing evidence around the effectiveness of early intervention in the case of victims of cardiac arrest, the UK ambulance service identified ways in which other people could be sent to victims believed to be in cardiac arrest. The adopted approach sees the attendance of a trained ambulance and crew supplemented by local responders, who range from off-duty medical staff to volunteer members of the public who receive the relevant training and equipment.

A number of UK fire and rescue services support the local co-responder approach and have developed the skills of firefighters in areas where a need has been identified by their ambulance service. Regular additional training is provided to fire and rescue staff by ambulance trainers, ensuring medical skills are maintained.

The role of local community first responders and fire and rescue service co-responders has continued to develop over time, with crews now able to respond to a range of medical emergencies.

ABOVE Equipment from the Incident Response Unit (IRU) deployed during a training exercise to deal with decontaminating casualties following a chemical spillage. *(Steve Greenaway)*

BELOW A FRS co-responder unit with their advanced life-support equipment and response vehicle. *(Hampshire FRS)*

Chapter Nine

Major incident case studies

A major incident is beyond the scope of business-as-usual operations, and is likely to involve serious harm, damage, disruption or risk to human life or welfare, essential services, the environment or national security. A major incident may involve a single-agency response, although it is more likely to require a multi-agency response, which may be in the form of multi-agency support to a lead responder.

OPPOSITE On 14 June 2017, the Grenfell Tower fire claimed the lives of 71 people. The incident was attended by over 200 firefighters and 40 appliances. *(Shutterstock)*

Buncefield Fuel Depot explosion and fire

On 11 December 2005, an explosion occurred at the Hertfordshire Oil Storage Terminal in Buncefield, Hertfordshire. The resulting fire became the largest fire to occur in Europe since the Second World War.

At the time of the incident, the oil storage facility at Buncefield was one of the largest operations of its kind in the United Kingdom. The facility was able to store approximately 60,000,000 gallons of fuel.

At 06:01 GMT on Sunday 11 December 2005, a massive explosion occurred followed by a number of fires involving storage tanks in the surrounding area. The subsequent investigation into the explosion and fire determined that vapour from leaking fuel from an overfilled tank ignited, causing a massive explosion. The explosion was heard up to 125 miles (200km) away from the incident, and such was the force of the explosion that it registered 2.4 on the Richter scale. There were subsequent explosions at 06:27 and 06:28.

The explosion led to damage and fires not just within the oil terminal, but also led to extensive damage to nearby buildings and caused fires in cars parked in the vicinity.

A mass evacuation was conducted by the police within the area due to the risk created by smoke, structural damage to buildings and the risk of further explosions.

Forty three people reported injuries as a result of the explosion although none were life-threatening. Fortunately, the timing and day of the fire meant that there were minimal people working in or around the site at the time of the explosion.

The flames from the fire were over 200ft (60m) high at its peak and the smoke plume could be seen over 70 miles (113km) away, while also being visible from space.

Following the receipt of what became many hundreds of calls to the incident, Hertfordshire Fire and Rescue Service mobilised around 150 firefighters to Buncefield.

The sheer scale and volume of the fires involving burning fuel meant that significant pre-planning and stockpiling of firefighting foam from all over the United Kingdom had

to be put in place before a sustainable fire attack could commence. Although an attack on the main fires in the fuel tanks had to be delayed, firefighting operations did take place to extinguish blazes in the surrounding areas and to provide protection to unaffected areas.

The detailed planning exercise prior to the commencement of an attack on the main fire had to take account of the amount of water and foam that would be required to extinguish the blaze and the subsequent potential for serious contamination of nearby rivers, streams and underground watercourses.

Operational resources sent to the scene included the use of a large number of National Resilience appliances and their crews from all over the country. Hertfordshire Fire and Rescue Service were assisted by 16 other UK fire and rescue services throughout the incident.

The operation to supply firefighting water involved the use of six high-volume pumps,

which extracted over 25,000 litres (5,500 gallons) of water per minute from a reservoir 1.5 miles (2.4km) away from the fire. The water supply was pumped through a system of hose lines, which incorporated a further six high-volume pumps positioned to boost the water supply to the incident.

There were over 20 fires involving burning fuel; firefighting operations to extinguish them commenced in the early hours of 12 December. At the peak of the operation over 32,000 litres (7,000 gallons) of firefighting foam was being pumped onto the fire per minute.

As a result of the sustained and pre-planned firefighting effort, over half of the fires were extinguished by 12:00 on 12 December. Despite the complex risks and nature of the fire, by 12:00 on 13 December just three fires continued to burn.

The final remaining fire was allowed to burn itself out under controlled conditions. Firefighters from Hertfordshire Fire and Rescue Service remained on scene for a number of days after the incident.

A government inquiry to determine the sequence of events leading to the explosion was conducted by the Health and Safety Executive and the Environment Agency.

The investigation found that a tank began to overflow at 05:20 on the morning of the explosion. A fault in an alarm system meant that it failed to communicate the issue. CCTV footage shows a large vapour cloud forming in the area of the overflow. It is estimated that this vapour cloud formed at ground level and was up to 2m (6ft 7in) in height.

The volume of the vapour cloud was sufficient to spread beyond the boundaries of the site. The damage and destruction resulting from the fire meant that it was not possible to determine a definitive ignition source, although the investigation did identify a range of potential causes.

ABOVE An aerial perspective of the devastation caused when the explosion ripped through the Buncefield Oil Terminal, Hemel Hempstead. It was extremely fortunate that nobody was killed. *(Getty Images)*

Boscastle flood

On 16 August 2004, a devastating flood swept through the small Cornish village of Boscastle. On the afternoon of 16 August, unusual weather patterns collided, leading to very heavy rainfall from localised storms on the hills above the village. The extreme rainfall entered two small rivers, which were quickly overwhelmed.

The speed and quantity of the downpour caused the two rivers to burst their banks. The two rivers converged and very quickly an estimated 2 billion litres (440 million gallons) of water flowed rapidly down the valley straight into Boscastle.

As a popular tourist destination, the village was busy with visitors and residents who had little time to react to the emerging torrent. The presence of large numbers of visitors made accurate accounting of missing persons difficult.

Such was the speed and weight of the deluge that buildings were damaged and destroyed and a car park was quickly overwhelmed, leading to more than 50 cars being swept out to sea.

Quick action by residents and visitors to move to places of relative safely and a subsequent rescue operation meant that, fortunately, nobody died.

Crews from Cornwall Fire and Rescue Service were among the first on scene, along with local volunteers from Her Majesty's Coastguard. The location of the incident adversely affected the Service's VHF radios and the mobile phone network was overwhelmed. The Fire and Rescue Incident Commander used a landline phone to send an assistance message, and throughout the following seven days crews from 29 of the service's 31 fire stations attended the scene.

Seven helicopters – three from the Royal Navy, three from the RAF and one from the coastguard – were scrambled to the scene in what is reported as the largest maritime rescue recorded.

Over 150 people were airlifted to safety with other people rescued by crews on the ground. The rescue operation continued well into the evening, with the subsequent search and rescue operation continuing for more than 72 hours. The extended period of rescue operation was established to account for missing persons and required the systematic check of damaged and silt-filled buildings as well as countless badly damaged and, in some cases, buried cars.

During the search and rescue operation, teams and resources were drawn from Cornwall Fire and Rescue Service, The New Dimension

BELOW The force of the rivers caused a torrent of water to devastate the village of Boscastle, Cornwall. *(Cornwall FRS, Wadebridge)*

ABOVE The force of the water carried vehicles down through the village and out to sea. A fire officer's response car can be seen in the foreground. *(Cornwall FRS, Wadebridge)*

Urban Search and Rescue Team from Devon (now Devon and Somerset) Fire and Rescue Service, specialist search teams from South Wales Fire and Rescue Service, the Marine and Coastguard Agency, Royal Navy, Royal Air Force, Royal National Lifeboat Institution, Devon and Cornwall Police, Yorkshire Police, North Cornwall District Council (now part of Cornwall Council) and the Environment Agency.

During the Boscastle flood 58 properties were flooded, four of which were completely destroyed. One hundred and fifty vehicles were swept away with 84 cars being recovered from the harbour. More than 30 vehicles were swept out to sea.

The clear-up operation extended over many weeks and required the removal of around 1,850 tonnes of debris.

Remedial work and flood relief work to prevent a similar future occurrence has seen over £10m invested in a wide range of flood defences. Organisational learning led to changes in safe systems of work for water incidents, along with enhanced provision of personal protective equipment.

LEFT As the water started to recede, firefighters were able to enter the flooded streets to undertake search and rescue operations alongside their multi agency partners. *(Cornwall FRS, Wadebridge)*

ABOVE The attacks on 7 July 2005 killed and injured many innocent people and challenged the combined resources of the emergency services. *(Getty Images)*

7/7 London terrorist attacks

On 7 July 2005, at approximately 08:50, bombs exploded on three London underground trains. An hour later, a fourth bomb exploded on the upper deck of a bus in Tavistock Square. The coordinated attacks were carried out by suicide bombers and represented the first known suicide attacks in Western Europe. Fifty-two people, including the four bombers, were killed in the attacks with over 700 people injured.

The bombs on the underground trains were detonated at approximately 08:50, shortly after the trains left Liverpool Street Station, Edgware Road Station and between King's Cross and Russell Square. The explosions occurred within minutes of each other.

The fourth bomb was detonated on the top floor of a double decker bus in Tavistock Square just under an hour after the first three.

Multiple calls were made to all three emergency services. The location of the trains between rail stations led to confusion over the location of the incidents and the number of explosions that had occurred.

The emergency response to the incident saw responders from all services working together to assist victims of the attacks.

Inquiries into the circumstances of the 7/7 bombings identified a number of issues that adversely affected the way in which emergency services were able to work together and communicate in the event of an incident of the scale of complexity encountered.

The implementation of the joint inter-agency interoperability protocols, advances in technology and other changes made across emergency services and other public bodies in response to this and other major incidents have been supported through lessons learned.

M5 motorway collision and fire

On 4 November 2011, at approximately 20:24, a collision occurred on the M5 motorway near Taunton in Somerset. The incident occurred in an area affected by a heavy

RIGHT The M5 collision and subsequent fire resulted in the death of seven people and had a profound effect on those involved and tasked with dealing with the tragedy. *(Getty Images)*

fog and involved 34 vehicles, consisting of 28 cars and vans and six lorries. Seven people died and 53 people were injured. The incident occurred on an elevated section of motorway just beyond the slip road from the Taunton junction.

On arrival, the first attending crews were faced with a rapidly developing fire involving multiple vehicles, which included articulated lorries and their cargo. There were people trapped in vehicles as well as casualties who had got out of affected vehicles either through the assistance of the public or under their own efforts.

The incident commander made a swift request for additional appliances and deployed those crews already on scene to rescue casualties and deal with fires where people were most at risk.

As additional resources arrived on scene, more crews were committed to assist their colleagues in what remained a developing situation. Further crews were then tasked to search vehicles and the surrounding area and to prioritise the extrication of casualties. With the quantity of water being used for firefighting, the incident commander instructed crews to seek a suitable water source.

In total, 21 fire appliances were mobilised to the incident during the fire and rescue stage, with resources coming from both the Devon and Somerset Fire and Rescue Service and neighbouring Avon. During the recovery phase, the Devon and Somerset Fire and Rescue Service Urban Search and Rescue Team attended.

Lessons learned from the incident included recommendations on rescue equipment, firefighting systems, special design features on motorways, recording of decisions and mobilising considerations. The incident highlighted areas of notable practice including the benefits of established procedures for the post-incident welfare of staff.

Grenfell Tower fire

On 14 June 2017 at 00:54, London Fire Brigade received the first of many hundreds of calls to a fire in Grenfell Tower, a residential tower block in Kensington, West London. Seventy one people died in the fire with a further 70 injured. Over 220 people either escaped the blaze or were rescued by firefighters.

Firefighters were faced with a rapidly developing fire of unprecedented scale. They worked in extremely challenging conditions to rescue those trapped by the fire and in their attempts to control the blaze.

Such was the ferocity of the fire that it took over 24 hours to bring it under control, and as well as the tragic loss of life the fire resulted in the destruction of 151 homes.

The fire investigation concluded that the devastating blaze was caused by a fire in a fridge freezer within a property on the fourth floor of the 24-storey building.

The initial fire in the building spread to involve exterior cladding that had been fitted to the building during a refurbishment that was completed in 2016. The fire developed rapidly.

London Fire Brigade mobilised over 200 firefighters and 40 fire appliances, with resources being sent from neighbouring fire and rescue services.

The incident has prompted a review of building regulations, which has recommended significant changes. A review of other tower blocks, both residential and commercial, has been carried out and remedial work is scheduled or has been completed where safety concerns have been identified. At the time of writing this book, the public inquiry remains ongoing.

BELOW The fire in Grenfell Tower developed rapidly, and firefighters faced conditions of the like they had never experienced before. *(Getty Images)*

FIRE

Chapter Ten

Fire investigation

Fire investigation requires an in-depth understanding of fire science and fire engineering. Like other investigative processes, it requires impartiality and a consideration of a range of facts, data and opinion. This section seeks to provide a broad outline of the purpose and process of fire investigation (it would require the whole book to even attempt to provide a detailed knowledge of this complex subject, which continues to challenge scientists and investigators in the UK and across the world).

OPPOSITE A Fire Investigation Officer (FIO) undertakes a detailed examination of a fire scene to ascertain the cause of fire and to look for evidence which may indicate arson. *(Devon and Somerset FRS)*

RIGHT This workshop has been devastated by fire and it will take many weeks before the business is able to operate again. *(Cornwall FRS)*

Fire investigation is an important activity to support organisational and industry learning and can directly contribute to the prevention of future fires. The fire and rescue service has been afforded investigative powers under the Fire and Rescue Services Act to perform this function.

Fire investigation is generally performed at two levels within the fire and rescue service. Simpler investigations will be completed by an incident commander, who will be required to determine and record the most likely origin and cause of the fire for national data recording. More complex investigation due to the nature of the incident or where the incident commander is unable to ascertain the most likely cause will be carried out by a specialist fire investigator, who will have undergone a greater level of training to perform the role.

In either of these situations, if criminal activity is suspected, the incident is immediately considered to be a crime scene, leading to an immediate request for police attendance.

The nature of a fire investigation will vary, dependent on the type, scale and severity of the incident. While the level of time and resource will be proportionate to the incident, the investigation will always be conducted using a structured and systematic approach to objectively gather information.

In all incidents, the person carrying out the fire investigation role will be required to maintain an impartial and open mind and to base any conclusion on facts and evidence. This requires consideration of a range of possibilities which are systematically dismissed in the absence of supporting evidence to determine a most likely cause.

The first stage of the investigation is conducted prior to any detailed physical examination of the fire scene. The purpose of this is to gather information from firefighters and other witnesses, CCTV, video and photographic footage, and other data sources such as fire detection systems and emergency 999 calls to the incident. This stage enables the investigator to build up a picture of the events leading up to and during the fire.

The next stage will involve an examination of the fire scene. The investigator will look at the extent of the fire, the fire behaviour and damage sustained. The damage and destruction caused by a fire can provide an array of information indicating the range of temperatures involved and the presence of an accelerant (a flammable liquid used to ignite, spread or increase the size and speed of the fire). Smoke and heat damage may show the direction in which the fire travelled and ultimately it can narrow down the point of origin, the place the fire started. Where this can be identified, the investigator can then start to consider potential ignition sources.

This stage will sometimes involve a detailed excavation of the fire scene. Sifting through the debris layer-by-layer enables the investigator to search for items that may be part of the ignition source. This careful process may require samples to be taken for analysis by a forensic scientist. Where the incident is considered to be a potential crime scene, this process will be carried out alongside a crime scene investigator. Where necessary, the fire investigator will reconstruct the fire scene with the remains of items in the room to gain an understanding of the fire development.

The final stage will involve considering all of the evidence gathered through each stage of the

RIGHT Fire development captured on CCTV. *(Freddie Martin)*

FAR LEFT The impact of heat damage at higher levels in a fire. *(Devon and Somerset FRS)*

LEFT This image clearly shows the importance of closing doors – the room behind the door has not suffered any fire damage. *(Devon and Somerset FRS)*

investigation. The investigator will have created a timeline of events using and applying the information gathered throughout the investigation. With a detailed understanding of the events and information gathered in the fire scene investigation, the fire investigator will be able to develop their hypothesis for the cause and origin of the fire. They will then systematically challenge this hypothesis using the information gathered in the investigation until they are able to support or dismiss this as the most likely cause.

In the event of a fatality, the role of the coroner is to determine the cause of death. The fire investigation will form an important part of the inquest process.

In the event of a loss incurring an insurance claim, the insurance company will often obtain information from the fire investigation and they may also carry out their own fire investigation.

In the event of a fire resulting from a criminal act, the police will lead on the investigation with the support of the fire investigator.

In each of these cases, where a case leads to either an inquest, inquiry or legal action, the fire investigator and, in some cases, attending firefighters, will be called to give evidence.

FIRE INVESTIGATION DOGS

A number of fire and rescue services have specially trained dogs that are able to identify a range of accelerants (ignitable substances). The smelling ability of a fire investigation dog is more accurate than any available technology in detecting accelerants. They can also assist with criminal investigations to determine whether a fire has been started deliberately.

The use of fire investigation dogs can significantly reduce the time required in fire investigation, and where there is no accelerant found this enables the fire investigation officer to focus on other potential causes.

Before being allowed to take part in an investigation, a fire investigation dog will be tested on a number of disciplines in a range of locations involving varying levels of challenge and difficulty.

Fire investigation dogs attend annual refresher training to maintain and test their skills in identifying substances correctly. They are trained to indicate a source and wait beside it until instructed by their handler.

Given the nature of a fire scene, fire investigation dogs wear special boots to protect their paws from sharp debris. They live with their respective handlers, who care for them both in and out of work time. Fire investigation dogs are transported in specially adapted vehicles driven by their handler.

LEFT Archie and Woody are fire investigation dogs in service with Cornwall FRS. *(Alfie Martin)*

Chapter Eleven

Firefighter welfare – post-trauma prevention and support

An increased understanding of mental health, and in particular the recognition of the increased risk to emergency service staff of the effects of post-trauma stress, has seen fire and rescue services develop preventative programmes and post-incident support to minimise the risk of firefighters suffering the debilitating and potentially life-changing effects of exposure to trauma.

OPPOSITE **The Fire Fighters Charity provide the Harcombe House Recovery Centre in the glorious Devon countryside as a dedicated site to support the mental health and social well-being of members of the fire service community**
(The Fire Fighters Charity)

RIGHT **Early intervention following critical incidents significantly improves the mental health recovery process.** *(Mary Pahlke)*

There are a number of approaches that are considered to be beneficial in minimising the risks associated with post-trauma stress and post-traumatic stress disorder (PTSD), which can become a serious condition if left untreated.

One such approach is the use of post-incident defusing, a model that uses trained peer supporters to debrief fire and rescue personnel who have attended what is deemed a critical incident. A critical incident will include those where someone is fatally injured, or they may perish as a result of injuries sustained. Other incidents such as the serious injury or death of a colleague, incidents with multiple trauma or incidents involving children or persons known to the crew are also deemed critical incidents. Where an incident meets the criteria for being classified as 'critical', attendance at a defusing session is mandatory. In addition, a manager or crew member can request a defusing session if there is a potential benefit.

A defusing session will be carried out as soon as possible after the conclusion of an incident and such is its value that it is deemed to be a part of the incident itself.

The aim is to build upon the resilience and coping that is core to a firefighter's skills. Understanding what a normal reaction is, what good coping mechanisms are, paying early attention to symptoms, identifying and attending to possible emotional injury to them and that of their colleagues and teaching awareness of what support is available is just some of the psych education.

This approach does not focus on those most at risk following exposure, as it has been the experience that firefighters have trained together, worked together, faced trauma together and thus work in crews, therefore intervention needs to mirror that. In addition, the aim is to intervene at the beginning. The aim is to make the psychological intervention as normal a part of the culture as the operational debrief or fire investigation.

Following a traumatic incident, the crew return to their station with their colleagues to be met by a peer supporter. They then participate in a group meeting that is about psycho-education, group 'story telling' combined with practical information aimed at normalising reactions to the traumatic event. It is a psychological first aid response. The peer supporter is also assessing whether a full psychological debrief, carried out by a trained psychological professional, is needed. Each meeting is monitored, actions taken outlined, risks noted and follow-up sessions organised. The participating group are also sent evaluation forms that assess how prepared for the session they were; the effectiveness of the introduction, the ability of the peer supporter, the firefighters' knowledge of

THE FIRE FIGHTERS CHARITY

The Fire Fighters Charity supports thousands of firefighters, serving and retired, and their immediate families through a broad range of services including rehabilitation programmes, health and wellbeing services, nursing support, recuperation and advice, information and support services to assist those in need.

The charity receives its funding through donations and fundraising activities and events, many of which are organised and run by firefighters and fire and rescue services.

RIGHT The charity was founded in 1943 as The Fire Services National Benevolent Fund (FSNBF) and changed its name to The Fire Fighters Charity in 2008. *(The Fire Fighters Charity)*

PTSD prior to the meeting and afterwards and how helpful the session was so they can give an overall evaluation of the meeting.

The peer supporter has undertaken training to equip them with a range of skills to be able to run the meeting, identify those more at risk, carry out a risk assessment, plan follow-up and refer to appropriate psychological support where required. Peer supporters are monitored and supervised alongside being required to attend regular training events in order to maintain their expertise.

The peer support process is underpinned by access to a professional counselling service to which individuals can self-refer where there is a need for a greater level of support.

BELOW Regular exercise is a great way to maintain mental health and it also provides opportunities to spend time with family and raise money for good causes. *(Brian Bracher)*

72%
62%
46%
19%
39%
61%
41%
11%
Price Avg 15 Vol Real-Time Weekly Volatility Index
20%
16%
18%
37%
Analytics
Audience
Daily Unique Sales by Country
Country
City
Status
Sex
Age
Interests
Collaboration
Behavior
Technology
Mobile
Custom 1
Custom 2
Custom 3
Income
Education
Daily Product Sales by Country

Chapter Twelve

Future firefighting

The development of computer systems, nanotechnology, robotics and sustainable energy sources present huge opportunities across all aspects of technology and the fire and rescue service will benefit both directly and indirectly from progress in global technologies.

Indirect benefits will come from safer homes, transport and materials. Direct benefits will come from advances in the technologies used in fire protection, communications, firefighting equipment and fire extinguishing media. Some of the advances being seen at present are explored in this chapter.

OPPOSITE Predictive analytics present opportunities to identify and deal with risks before an incident occurs. *(Shutterstock)*

ABOVE The Rosenbauer concept fire truck (CFT). *(Rosenbauer)*

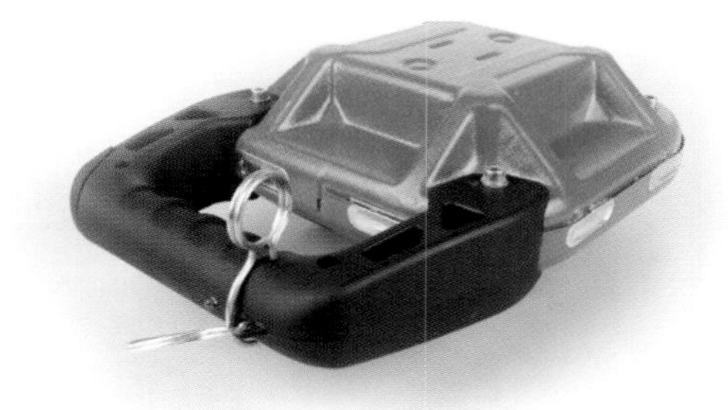

RIGHT GreenSol® active fire suppression systems – fire extinguishing device using nanotechnology. *(Clarita Solutions Ltd)*

Predictive analytics

Predictive analytics is the term used to describe the use of data from multiple and diverse sources, along with statistical algorithms and computer software (including artificial intelligence) to both identify and predict future outcomes.

Within the fire and rescue service this presents wide-ranging opportunities to predict future incident trends and to address them at source, stopping something before it can start. Currently data, held in isolation, has a limited ability to help services understand future trends. With advances in predictive analytics, wider sharing of data and the volume of available data, there is a genuine opportunity to not only learn from the past but to use the past to create a safer future.

Electric-powered fire appliances

With the ever-increasing development of battery technology, the demand for sustainable fuels and a commitment by the UK government to rapidly reduce vehicle emissions is forcing the development of electric vehicles.

European fire appliance manufacturer Rosenbauer have already developed and built an electric-powered fire appliance. While this remains a prototype it is proving the technology and there are plans to produce a vehicle based on the concept. Other manufacturers are already following the lead taken by Rosenbauer.

Under the project name of the concept fire truck (CFT), Rosenbauer have widened the use of technology so that the CFT isn't just an electric version of diesel fire appliances in use and with the agility that electric power presents, it is expected that future fire appliances will both drive and operate pumps with increased performance capabilities.

Nanotechnology

Nanotechnology is the term used to describe the science of manipulating and engineering matter at the molecular and atomic level. The manipulation of molecules enables changes to and the creation of some really interesting materials that are capable of achieving new things or vastly improving properties of current substances and materials compared to what they do now.

The fire and rescue service are already seeing some of the opportunities and benefits resulting from improvements to materials used in the manufacture of firefighting clothing.

Nanotechnology has also been used to produce fire extinguishing adhesive plates with component parts that automatically activate when a pre-determined temperature is reached. At this point the plates decompose, releasing a gas capable of extinguishing a fire and preventing fire spread.

The fire extinguishing plates can be manufactured in very small sizes to protect even the most difficult places to access or cover with conventional firefighting techniques.

Drones

Drones, also known as unmanned aerial vehicles or UAVs, have become popular for a range of uses. A number of fire and rescue services already use drones for aerial reconnaissance to capture images that assist in directing resources and to monitor fire spread, with both live footage and enhanced imaging using thermal image cameras.

Drones are now available with the capability to carry a payload and plans are in place to create a medical response capability through transporting defibrillators. A rescue drone has also been created to transport a liferaft to a person in water at sea. The future use of drones in fire and rescue holds exciting prospects.

ABOVE The use of drones in the fire and rescue service is becoming commonplace. *(Jason Andrews and NESTA)*

Radio Frequency Identification tracking

A Radio Frequency IDentification (RFID) tracking system is able to transmit data from an RFID tag to a remote receiver. The data it can transmit can be as simple as sending a location, useful for keeping track of a breathing apparatus wearer within a smoke-filled property, but it can also send system data from an inbuilt transmitter linked to a computer. This then allows active monitoring of any data that the computer is monitoring. In the case of a breathing apparatus wearer, this could include the body temperature of the wearer, the amount of air being consumed and information on any distress signal. In this case, the transmitter may just be of a relatively short range with a direct link from transmitter to receiver.

When an RFID is linked to a data network, including mobile phone networks, the range of transmission is global. A practical example of this is a fire appliance manufacturer installing RFID technology within the fire appliance management system. In this case, the data can be monitored 24 hours a day from any location, enabling not only active monitoring of performance but also enabling the rapid resolution of a fault if one should occur. This is particularly important for something as risk critical as a fire appliance.

LEFT RFID tracking technology can assist in accountability and equipment availability and servicing. *(Shutterstock)*

Powered exoskeletons

An exoskeleton is a powered and articulated frame worn by an operator to work in tandem with the wearer. The frame enhances the human wearer's performance through physical strength and acts to protect the wearer from external forces, for example if carrying a heavy load.

Exoskeletons could be used by firefighters in a range of situations, such as carrying heavy firefighting equipment long distances or when ascending a tall building where use of the lift is not achievable.

The use of powered joints and a rigid frame would remove the physical strain on the wearer while leaving them in full control of movement and decisions.

LEFT An exoskeleton designed for firefighting use. *(TRIGEN Automotive)*

Appendix One

Support services

In addition to operational activity, a fire and rescue service requires support functions in order to deliver its services efficiently and effectively. Non-operational roles in the fire and rescue service are wide and varied, requiring a range of professional skills. In some services, a number of support functions may be shared or outsourced to other organisations or companies.

Support services commonly seen within the fire and rescue service will include:

- Human resources
- Pay and conditions
- Administration services
- Corporate communications
- Fleet management and workshops
- Procurement
- Operational planning, research and development
- Information, communication and technology (ICT)
- Health and safety
- Hydrant maintenance
- Media
- Data analysis
- Fire authority member support

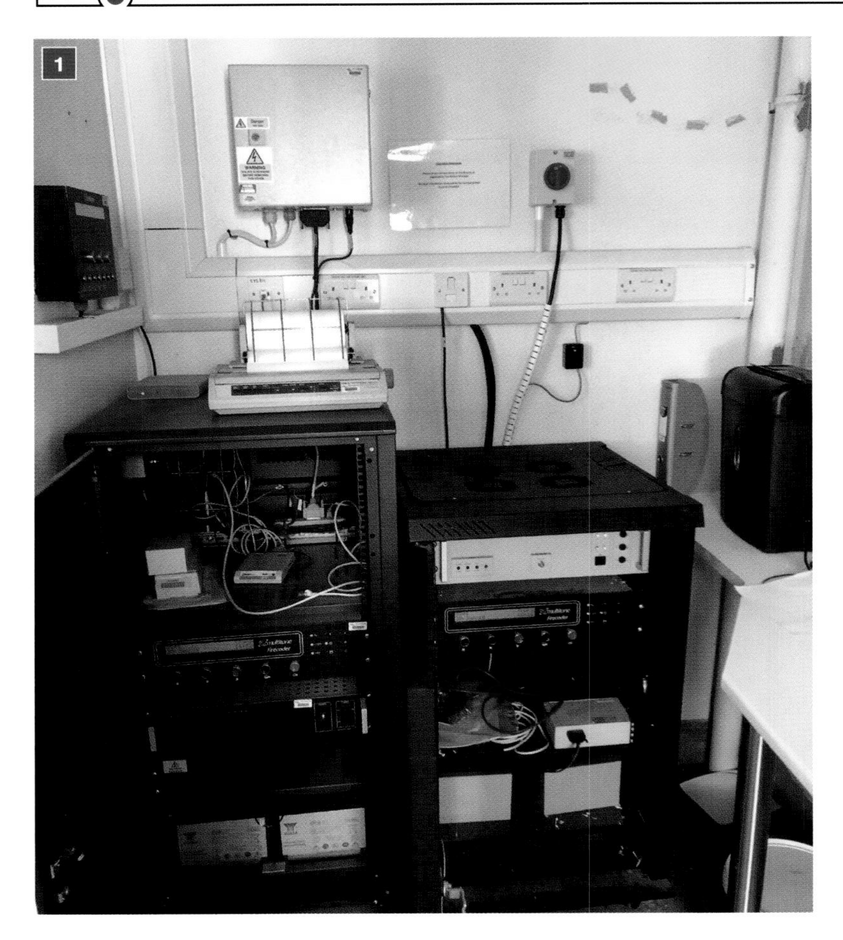

1 ICT equipment used to active firefighters pagers, notifying them of an incident and for receiving information about the type of call they are attending. *(Cornwall FRS)*

2 A maintenance technician testing a fire hydrant. *(Devon and Somerset FRS)*

3 **The fire and rescue service workshops are responsible for the maintenance and repair of the vehicles and equipment used by firefighters. Highly skilled technicians ensure that fire appliances remain fully operational and can respond when required.** *(Cornwall FRS/Callum Matthews)*

4 **A fire officer taking part in a media interview at the scene of a flooding incident. Providing information to the public during an emergency is important and key officers receive training to undertake this role.** *(Cornwall FRS)*

5 **Interpretation of data ensures that resources and activities are provided at the point of greatest need.** *(Shutterstock)*

Appendix Two

Model recruitment guidance

The wholetime recruitment process

Anyone aged 18 or over can apply to become a firefighter; but candidates must have the potential to meet all the requirements of the role.

There are a range of duties carried out by a firefighter, but a lot of the job is based around working within the community, so good communication skills are essential. There is also the physical element of the role, such as using ladders, hoses and other equipment; therefore, candidates are required to reach a prescribed standard of physical fitness.

There are various stages of the application process and each candidate must complete and be successful at each stage to progress to the next one. If the service does not feel that the candidate is suitable at any stage of the process, the candidate will be advised that they will not progress any further. The candidate may therefore be unsuccessful at any stage in the process detailed below.

Stage 1 – Application form

In the application form each candidate will be asked a series of questions. These will help the service to decide whether the candidate is likely to be suitable to become a firefighter and whether they are eligible to progress.

- Equal Opportunities Monitoring Information – this information will not be used as part of the selection process. It is for monitoring purposes only.
- At the application stage, under the Rehabilitation of Offenders Act 1974, candidates are obliged to declare if they have any unspent convictions or criminal proceedings pending as part of their application as these may bar them from working for the service.
- Any personal data that the service collects during the process will be destroyed should the candidate not be successful.
- If a candidate provides false information their application will be rejected.

Stage 2 – Online assessments

If the candidate is successful at Stage 1, they will be sent two online assessments, by email, for completion by a specified date. They must complete the assessments by this date. These assessments will measure the candidate's suitability for the role based on personality characteristics and aptitude, as well as assessing their problem-solving ability.

Stage 3 – Job-related assessments

These are exercises designed to measure practical and physical ability. This will include assessments on:

- Ladder climb
- Casualty evacuation
- Ladder lift
- Enclosed space
- Equipment assembly
- Equipment carry
- Aptitude assessment

Upon successful completion of Stage 3, the candidate will be invited to an assessment centre.

Stage 4 – Assessment centre

This will be approximately half a day and will include:

- Formal interview and presentation
- Group discussion with other candidates

Fire deaths 39, 128-129, 151, 155, 158
Fire detectors 43-44
automatic 43-44
heat 44
ionisation 44
optical 44
smoke detectors 43-44, 128-129
Fire engines – see Fire appliances
Fire extinguishers and extinguishing 33-35, 84-91, 161
automatic systems 48
carbon dioxide 34, 90
colour codes and use 89
dry powder 34, 90
foam 34,84-8890, 107
nanotechnology devices 162
removing fuel, heat and oxygen 33
very high-pressure cutting systems 84, 105
water 89
wet chemical 90
Firefighter safety 52
incident command system 52
working at height regulations 75
Firefighters 20
fatalities 128
incident commander 123-124, 154
on-call 123, 167
post-incident welfare 151, 157-159
recruitment guidance 166-167
safety 129
volunteer 18
Fire Fighters Charity 159
Fire investigation 7, 153-155
determining point f origin 154
examining the fire scene 154
fire investigation dogs 155
fire investigation officers 153-155
gathering information 154
hypothesis for the cause 155
Fire legislation 10
Firemarks 11
Fire prevention 12, 16, 36, 39-42, 50
education 39-40
home safety visits 40
maintaining closed doors 155
safety education 6, 128
Fire protection and precautions 36, 41-42, 50, 129, 161
active 43
fire resistance 42
passive 42
Fire, science of 30, 153
fuel 30-31, 33
heat 30, 33
oxygen 30, 33
Fire Scotland Act 2005 23
Fire Service College, Morton in Marsh 38
Fire Services Act 1947 12
Fire Services in Defence 22
Fire station 6, 21, 57-63
appliance room 58-59
crewing arrangement 58
drill yard 61-62
fitness room 60
forecourt and road access 62
red flashing l ights 62
kitchen dining area 60
muster bay 59-60
office 59
outside training area 58
pole drop 60
study and rest rooms 61
training/meeting room 59
training tower 61
tri-service stations 63
Fire stations
Hayle Tri-Service Station 63
Leadhills Community Fire Station 18
River Thames pontoon 114
Fire suppression systems 45, 49-50, 129
domestic 49-50, 129
sprinklers 43, 45-49, 129
water-misting 129
Fires and incidents 127-143
airports and aircraft 10-108, 135-136
basements 130
commercial buildings 7, 107, 128, 130
compartments 51, 84
construction and demolition sites 36, 131-132
dwellings 128-129
chip pans 129
white goods 129
factories 132-133
farms 136
food processing and packaging facilities 133
fuel 35, 84, 139, 146-147
hazardous materials 142-143
heritage buildings 137
high-rise buildings 81, 129, 151, 155, 163
external cladding 129, 151
house fires 7, 17, 49-50, 128
landfill sites 133
London bridge 1212 10
offices 130
railway stations and infrastructure 134-135
recycling sites 134
rural 136-138
ships 135
shops and shopping centres 130
thatched cottages 33
transport hubs 134
warehouses 131
wildfires 137-138
workshops 127, 132-133, 154
Foam making equipment 86-88
compressed air foam system (CAFS) 87, 105

Great Fire of London 1666 10-11

Hampshire Fire and Rescue Service 102, 106, 113
Health and Safety Executive (HSE) 147
Heat transfer 31-32
conduction 31
convection 32
fire spread 31
radiation 32
Helicopters 112, 148
Her Majesty's Coroner 7, 155
Hertfordshire Fire and Rescue Service 146
HM Coastguard 148
Home Office 12
Hoses and fittings28, 79-81, 112, 117
branch 28-30, 83
Cleveland coil stowing 81
connections 28
controls and fittings 79, 83
couplings 80
Dutch rolled hose 80-81
flat laid delivery hose 28
high-pressure hose reels 79-80
remote-controlled nozzles 107
running out 29
suction hoses 28, 81, 83
Hydrants 28-29, 78-79, 112, 164
marker posts 79

Incident handover 124
Incident logs 124-125
Insurance claims 118, 155
Insurance Fire Brigades 11
Integrated Risk Management Plan (IRMP) 26
Iraq 22

Joining the FRS 16
Joint Emergency Services Interoperability Protocols (JESIP) 52-55
METHANE messages structure 54
working with other agencies 55

King Charles II 11

Ladders 29-30, 73-75, 109
manual 29
pitching 29
roof 74-75
triple extension 74
turntable 108-109
10.5-metre 74, 103
13.5 metre 29, 74, 103
Lancashire Fire and Rescue Service 107
Lighting, emergency 43, 129
Local Authority Fire Services 11-12, 22
National Occupational Standards 22
London Fire Brigade 53, 114, 125, 151
London Fire Establishment 11

Major incidents 53, 65, 116-118, 125, 145-155
Boscastle flood 148-149
Buncefield Fuel Depot explosion and fire 146-147
Grenfell Tower fire 151
large fires 119, 146-147, 150-151
members of the public affected 118
missing persons searches 119
M5 motorway collision and fire 150-151
volunteer staff 118-119
weather-related events 119
widespread flooding 119, 141, 148-149
7/7 London terrorist attacks 150
Marine and Coastguard Agency 149
Medical equipment 98
defibrillators 98, 163
medical response kits 98
resuscitation equipment 98
stretchers 98
Ministry of Defence 21-22

National Emergency Services Network (ESN) 112
National Fire Chiefs Council (NFCC) 114
National Fire Service 12, 28
National Health Service (NHS) 143
Emergency Medical Service 143

National Resilience Programme 114-117, 125
NATO 22
North Cornwall District Council 149
Northamptonshire Fire and Rescue Service 112
Northern Ireland 23, 114, 141

Officers' response cars 18

Personal protective equipment (PPE) 68-71, 111
automatic distress signal unit (ADSU) 70-71
boots 69
breathing apparatus 52, 69-71
personal line 70-71
tally 72
compartment firefighting kit 66-67
gas-tight suits (GTS) 68-69
gloves 69
helmets 19, 69
technical rescue kit 67-68
torches 71
water rescue kit 68
Police 7, 11, 55, 63, 112, 149, 154-155
police community support officers (PCSO) 63
Police and Fire Reform (Scotland) Act 2012 23
Positive pressure ventilation (PPV) 90-91
Post-trauma stress disorder (PTSD) 158-159
peer supporters 158-159
post-incident defusing 158
Pumps and pumping 27-29, 75-78, 116-117
ejector pumps 78
hand-drawn 102
horse-drawn 102
impeller 77
intermediate pumps 29
light portable pumps 77, 83
major pumps 75-76
multi pump 74
primer 76-77
specialist pumps 78
steam powered 102

Queen's Flight, The 22

Rescue and extrication – see also Technical rescue 91-97
airbags 94-95, 110
air tools 110
animals 142
at height 108-110
below ground 109-110, 134
combination tools 92-93
confined spaces 110-111
glass management 96, 140
hydraulic tools (cutting and spreading) 91-93, 97, 105, 110, 139
jacks, blocks and wedges 110
off-road terrain 106, 111
reciprocating saws 93
rescue lines and harnesses 111, 141
rescue platforms 96-97
rescue rams 93
road traffic collisions 97, 110, 138-140, 150-151
team approach 139-140
sharp end protection 94
small gear hand tools 98, 110
stab fast jacks 95-96
stabilisation 95-96
trapped persons 115, 139-140
immobilisation and extrication 140
vehicle airbag covers 98
Response 50-51, 143
multi-agency 135
Roles 18-21
area manger 20
brigade manager 21
crew manager 20
firefighter 20
group manager 20
station manager 20
watch manager 20
Romans 9-10
Emperor Augustus 10
Royal Air Force (RAF) 148-149
Royal Air Force firefighters 22
Royal Berkshire Fire and Rescue Service 102
Royal National Lifeboat Institution (RNLI) 141, 149
Royal Navy 22-23, 149
aircraft handlers 22-23
firefighters 23
Fleet Air Arm 23
Naval Air Command Fire & Rescue Service 23
Royal Naval Air Stations 23
RSPCA 142

Salvation Army emergency response service 118-119
Scotland 11-12, 18, 23, 114, 141
Scottish Fire and Rescue Service 38, 105
National Training Centre, Cambuslang 38
Second World War 11-12, 138
safety poster 12
the Blitz 9, 12
Smoke 31
different colours 31
South Wales Fire and Rescue Service 149
Specialist training centres 38
Spontaneous ignition and combustion 137
Sprinklers – see Fire suppression systems
Support services 164-165
Surrey Fire and Rescue Service 111

Technical rescue – see also Rescue and extrication 10, 138-142
evaluation 140
full access 140
safety and scene assessment 139
space creation and access 140
specialist line and confined space rescue 141
stability and initial assessment 140, 142
Thermal image cameras 70, 73, 107
Training 11, 25-29, 35-39, 51, 58-59, 61, 105, 159
acquisition of skills 29, 36
classroom 30
driver training 37
extinguisher 89
fire investigation 154
initial skills (basic) 6, 27-29, 36
ladders 74
multi-agency exercises 53
ongoing and development 38
phases 36
PPV 90
specialist skills 37-38
standard drill 27
standardised 12
theoretical knowledge 29-30, 39
water rescue 38
working at height 29
Tri-service safety officers (TSSO) 63

Urban Search and Rescue (USAR) 115-116, 141-142, 148-149
Modules One to Five 115-116

Wales 12, 23, 114, 141
Water rescues 38, 111, 141
inland flooding 111, 116, 141, 148-139
inland water courses and lakes 111
life-raft delivered by drone 163
motorised boats 111
inflatable paths 111
Water supplies 27-29, 78-79
collapsible dams 83, 112
controlling under pressure 30
delivery 28
open water sources 27-28, 79, 112, 116, 147
supply 28, 116
fixed supply systems 129
Welfare support units 13
West Sussex Fire and Rescue Authority 26
Works fire services 23

York, Duke of 11

Bibliography

Elementary Fire Engineering – Institution of Fire Engineers (2007)
Firefighters Handbook (Essentials of Firefighting and Emergency Response) – Delmar Thomson Learning (2004)
Fireman's World – Anne Burridge (1983)
JESIP (Joint Emergency Services Interoperability Principles) – Internet (2019)
UK FRS National Operational Guidance – Internet (2019)
Post Incident Support - Vanessa Davies - International Firefighter Magazine (2014)
Wikipedia (2018)